THE
# FARMER'S
# WIFE

# HELEN REBANKS

# THE
# FARMER'S
# WIFE

## My Life in Days

faber

First published in 2023
by Faber & Faber Limited
The Bindery, 51 Hatton Garden
London EC1N 8HN

Typeset by Faber & Faber Limited
Printed in the UK by CPI Group (UK) Ltd, Croydon, CR0 4YY

A CIP record for this book
is available from the British Library

ISBN 978–0–571–37058–0

MIX
Paper | Supporting
responsible forestry
FSC
www.fsc.org
FSC® C171272

Printed and bound in the UK on FSC® certified paper in line with our continuing
commitment to ethical business practices, sustainability and the environment.
**For further information see faber.co.uk/environmental-policy**

2  4  6  8  10  9  7  5  3  1

For James, Molly, Bea, Isaac and Tom

———————

'But the effect of her being on those around her was incalculably diffusive: for the growing good of the world is partly dependant on unhistoric acts; and that things are not so ill with you and me as they might have been, is half owing to the number who lived faithfully a hidden life, and rest in unvisited tombs.'

George Eliot, *Middlemarch*

# CONTENTS

# RECIPES

EXTRA RECIPES

# DAWN

The cockerel crows. 5.30 a.m. I pull the blanket over my head, trying to hold on to the night, just a little longer. Some days there is a blurry moment just before I wake up, when I exist in a dreamlike state. I forget which bit of my life this is, forget that I am a mother and a wife and that I have a thousand things to do. I didn't always have these roles, but I knew them well. I grew up in a busy farmhouse. My bedroom was in the attic. Some mornings I would lie staring through the skylight to the clouds, with a headful of teenage ideas about how I'd escape the farm. The noises of the kitchen would drift up the stairs. The kettle boiling. Dogs barking. Doors banging. Mum calling for help with the work or for me to get ready for school. I dreamed of being an artist and travelling, with days that opened out before me to create things, and with lots of time to think and read. I didn't want the life of a farmer's wife. The women and girls worked indoors and smelled of soap. Their chores never ended – washing, ironing, cooking and cleaning. Men and boys did the outside work; they smelled of muck. They lived by a dirty, wet and cold routine of milking, feeding and shepherding, and didn't talk about much else. I hated the bind of 'the farm'.

But, despite all my girlish ideas, I am now here, in my own farmhouse on a hillside in the Lake District, just six miles from where I grew up. I live with my husband, James, and we have four children – Molly, Bea, Isaac and Tom. There are also six sheep-dogs, two ponies, twenty chickens, five hundred sheep and fifty cattle to care for. I am a farmer's wife, and this is my story.

My dad often says, 'You make your bed, you lie in it.' I recoil every time I hear it, usually because he says it when I am struggling with something. I don't find it kind or helpful. I know what he means – that we all live by our choices and they have their costs. It is kind of true, we can't 'have it all'. But that hard old saying doesn't offer any possibility for change. It suggests that a bed, or a life, is made once and is then fixed like that forever. It suggests that you can't ever grow and change but must simply suffer and endure. But I think we make our beds anew every day – life is really a constant remaking and reshaping of ourselves and our days. I am always looking for different ways to 'make my bed' and for ways to avoid becoming stuck.

A fly is buzzing at the window. I get up and open the latch to let it out. A cuckoo calls across the green valley. James has already gone out to check on a cow calving and the rest of the house is asleep, despite the racket outside.

I wrap my pale-blue dressing gown around me and carry three cups downstairs that the children have abandoned. There are tell-tale crumbs of stolen biscuits on the carpet.

I take my stainless-steel kettle to the sink, tip it out and refill it, light the gas on the stove and put it on to boil. I circle the kitchen, the room we live, work, cook and eat in. After shaking the cushions back into shape, I place them neatly on the grey velvet sofa. I pick up toy dinosaurs and discarded socks from the floor and tidy a pile of papers and yesterday's post. The flowers from our garden have wilted so I gather them up and put them out, setting my grandma's old vase in the sink to wash later. It would have been her birthday today. I wipe the table and straighten the wooden chairs around it. The flagstones are cold under my bare feet, so I find my slippers. Floss, our retired collie, is still in her bed with no desire to get up so early. She wags her tail at me gently as I pat her. I make some tea in my favourite mug and clutch it in two hands, the tea steaming my face. I go and sit snug on my nursing chair. I have kept it near the fireplace, beside the arched double doors that lead onto the patio in front of the house. I no longer need this chair, but it is my cosy place in our busy family home. Rocking back and forth, I am nostalgic for the days and nights that I nursed my babies, but relieved too that I no longer have a little one clamped to my breast or needing to be rocked to sleep through the restless stages. I am through those all-consuming baby days and the best memories are warmly tucked in my heart.

The sun's rays pierce through the leaves in the trees and a red squirrel hops along the top of the garden wall, oblivious of being watched. I pick a book up from the footstool in the window and try to read, but my mind wanders across the words on the page.

I love this quiet time in the house. Soon it will be noisy and chaotic.

Upstairs, Tom calls out 'Mum' but then rolls over and has gone back to sleep by the time I reach his bedroom. I take a bundle of dirty clothes from the landing, sort them and put a load in the machine. I hang wet things from the previous wash on the rack and fold a basketful of dry clothes. Through the glass door of the utility room I can see our two saddleback pigs snoozing among the tussocky grass. They have been rooting around a patch of ground that I want to grow vegetables in next year. They flick their ears and lie on their sides, nestled into each other.

The familiar sound of a lamb bleating makes me finish sorting the clothes quickly. I have been feeding her in the sheep shed for a week or so, as her mother was poorly and didn't have much milk. Her mother has now recovered and they are living back outside in the field next to the lane, but she doesn't have enough milk yet so I am still helping. The lamb now thinks she has two mothers. I go to the kitchen and mix powdered lamb milk from a tub near the sink with warm water, and put it into a bottle with a rubber teat. She has squeezed through the garden gate and waits noisily by the kitchen door.

I sit outside on the wooden bench feeding her in my slippers and listen to the birds chirping around me. She takes no time at all to suck the bottle dry and nearly knocks it out of my hands before racing off to find her slightly cross mother who is waiting for her. The blush-pink roses are blooming against the blue Lakeland stone of our house. As I walk down the garden steps Floss follows me. The lavender brushes against me from the raised beds by the gate. I am soon down in the paddock in front of our house, stamping on the nettles and dockens with my slippers to make a path. I let the hens out of their coop, check for eggs in the nest boxes and load up my dressing-gown pockets with five deep-brown-coloured eggs. I dip the hens' water bucket into the beck and hold it under the flow to fill it. Back at the coop, the hens run and gobble up water in their beaks. They tip their heads back to let it flow down their throats. Floss sniffs the ground; there has likely been a fox here through the night. The hens cluck around me and I throw a little of their feed on the ground as I have no scraps of kitchen waste with me right now. I am always feeding or watering someone or something.

The fresh morning breeze in my hair feels good. I tell myself that I mustn't stay indoors all day. It is easy to bury myself in the housework, doing everything for everyone else. It is up to me to share the load. The children need to learn to do some of the jobs themselves. I have to make my family see the unseen jobs and value them.

Sometimes I only see the pile of dirty washing on the floor and lose sight of the ever-changing world outside. Visitors tend to think our life is idyllic because they turn up on a sunny day, when all is green and the valley looks stunning, and we are all smiling, but we are much like any other family: we work hard to pay our bills and our individual moods change like the weather.

An hour later, everyone else is up and the house is humming with electric toothbrushes and chatter. The quad bike roars into the driveway and Bea jumps off it and runs in to reluctantly swap her farm boots for school shoes. She has been up to the barn to feed

Bess's litter of puppies – they will be off to their new homes next weekend. I manage to brush the hay off the back of her jumper as she walks to the car. 'Quick, you'll be late for the bus,' I say. I pass James half my slice of toast as he turns and follows her to the car. 'I'll cook you some bacon. It'll be ready when you get back.' I can hear the girls arguing about who left the wheelbarrow full of muck and didn't tip it out. They have to try and switch from being at home, working on the farm, to being stuck in a classroom all day, and I know which they would prefer. Molly is about to start her exams and can't wait to leave school. As far as I can tell, nothing is inspiring her to stay there.

I chase Tom around the kitchen to get him dressed. 'I need to put these on you so you can go to nursery – you can't go in your pyjamas!' He giggles as I grab and tickle him. 'I want to stay at home and play with my dinosaurs,' he says. 'It's OK, you'll be back at lunchtime, it's only a morning today. Granny will pick you up and bring you home. I promise I won't touch your game on the floor. It will be exactly the same when you come back.' He twists his body into all sorts of awkward shapes to make it more diffi-cult for me to dress him. Once his socks are on, he does a wiggly half-naked dance and is now determined to do the rest himself. He gets tangled with his long sleeves by trying to put his arm through the head hole of his top.

After helping him, I crack three eggs into a bowl, melt some butter in the frying pan and stir and fold the eggs with a wooden spoon. Isaac is busy buttering his own toast. 'Put a slice in for me, please,' I say, and spoon some of the egg that I have cooked onto a little plate for him. 'Here, eat this, it will fill you up.' Tom sits up at the table when I get him a bowl of yogurt and his favourite granola with the 'pink bits' in.

The boys' primary school is a fifteen-minute drive along the lake road. I spend hours driving this route. On days that I find it a chore, I make myself remember that it is one of the most scenic roads in

Britain. The fells open out as we leave the village, and the lake is sparkling in the sunshine down below us. Tom presses the button on the car window as we curl down the road; he tells us that he wants to feel 'the wind of change' on his face and makes me and Isaac laugh with his nonsense. I can see his chubby fingers gripping the window as it doesn't go all the way down, his little nose scrunched up against the glass. I ask the boys to give me one word to describe the water today – Tom says 'deep' and Isaac says 'glistening'.

I know this road so well. I know its blind spots, where I need to slow down and where I can speed up. Driving it twice or three times a day has made it become muscle memory. I watch out for Tommy on his tractor or Adam in his pickup as I go around the huge lump of rock that juts out on a particular bend. When we get to school, just in time, the boys run off to their classes without looking back. 'Love you . . . See you later!' I call to the wind. As I turn back to the car, I feel the strange mix of relief, freedom and crushing emptiness that every mum knows.

As I drive home, I pass people standing on the shore taking photographs. One couple are unloading a kayak from the top of their car. The man who feeds the swans every day in Glencoyne Bay is out, knee-deep on his paddleboard. The white birds flock around him. The water looks serene. I drive on and spot a space in a lay-by near one of my favourite places, so I pull over. The ground beneath the trees is covered in a carpet of bluebells stretching from the road to the shoreline. I walk through their delicate flowers. Just a few weeks ago the bluebells were dormant and hidden beneath the frozen ground. They spent the spring soaking up energy from the sun through their leaves and now they have burst forth and bloom. And when they finish flowering they will die back into their bulbs and repeat the cycle. I love that something so small and pretty can be so tough.

I walk down and sit on a rock by the shore. I look across Ullswater to the little rocky bays and the scrub on Hallin Fell. It is so peaceful.

The holiday season has only just begun. The water is still too cold for tourists to splash about in, the wind too sharp for having barbecues on the shore. The lake laps gently around the stones by my feet. The branches above me burst with vivid green shoots and leaves.

I can afford this pause, a moment for reflection on my own, because we have just got through our busy lambing season. It has been an intense six weeks of teamwork during which our own needs came second to the sheep. Lambing time brings our children right up close to life and death. I can already see in our three older children a deep understanding of what looking after livestock and land means. They know the world is bigger than any one of us, that nature carries us forward.

My phone rings and I see the accountant's number. He is chasing me for some forms we were supposed to have signed and returned. I tuck the phone back in my pocket, guilty that I am stealing time. I will ring him when I get home.

I drive back to the farm, up the windy road past Gowbarrow Fell, and through the village, and pull into our lane. A van is pulling out. 'Is your husband about?' the driver says to me.

I look at him for a second. He is wearing a shirt and tie, with a company logo on the sleeve of his jacket. 'Why?' I ask.

'Well, I wanted to talk to him about road planings for your lane – I can get him a good deal.'

'How much a ton are they?' I ask, my tone irritated because he wants to speak to my husband instead of me. I can see James in the distance on the tractor, lifting a hay rack back to the yard. He hates salesmen. The man in the van is surprised by my question, and looks to a clipboard with notes on it, fumbling towards an answer. But before he gives me a price I say, 'The last lot that I bought were rubbish', adding, 'Far too much dust in them – and, look, they haven't filled in the potholes very well.'

'I'll leave you my number,' he says quickly, and I take his card through my window and chuck it in the footwell.

I drive up the lane to the house. The hens are pecking around in the raised beds for slugs, and Floss is waiting by the door for me.

I am never quite sure what kind of day I am going to get. Some days, all I can do is firefight. I am pushed to and fro by events, the weather, or the needs of the farm animals or my family. Sometimes it seems relentless, but I try to approach every day as a new start and as more than just chores. I look for the beauty in the world around me. I try to learn something new every day. And I remember that this busy life we have created grew out of the love we have for each other. I know that we can do hard things.

# 1

The big square farmhouse sits up on a hill overlooking the local town. Surrounded by 120 acres of good land with deep-red soil. My grandparents came here in 1946, signing the tenancy just before they married. Grandad was a tall, broad-shouldered man, always dressed in a plain shirt and a jacket, and grey trousers held up by braces either side of his huge belly. I never saw him in dirty clothes. Farm men of that time earned their status, and the respect of their peers, through good 'stockmanship', a thing everyone knew when they saw it, even if they rarely explained what it was. It was seen in the shine on your horses' coats, and the size and health of your cattle, and the beauty of your ewes on sale day. Grandad was a decent farmer, but above all he was a horseman. He was known far and wide for breeding pedigree prizewinning Clydesdale horses. On Thursdays my grandma cleaned the silver in the cabinet in the sitting room. It was full of trophies, including one for winning the male champion at the Royal Highland Show with a horse called Bell Mount Ideal. Once, a man arrived at the railway station in town and walked all the way to the farm with a suitcase full of cash. He bought a mare to go back with him to New Zealand. In the 1950s and 1960s Grandad exported horses to Canada. He travelled to the Winter Fair in Toronto to watch teams pulling carriages around the arena.

Grandma always told us about the distant places they had been to together, but she sounded sad that they never actually visited the cities or towns or saw any of the sights. Their trips were always straight to the shows or livestock auctions and back home, only stopping in a hotel if absolutely necessary.

Grandma was one of eight children and she grew up on a small village farm where they worked the land with horses. She remembered running to the next village to fetch the doctor when her mother was giving birth to her youngest sister. The doctor arrived in his horse and carriage to assist the birth, just in time to save the baby. My great-grandmother was small and wore her grey hair in a bun; she has little circular-framed glasses in all the old black-and-white photographs. The children all helped run the family farm, milking the cows, making butter and cheese, hoeing and harvesting potatoes, washing and cooking. Grandma dreamed of being a teacher, but it wasn't to be. She seems to have settled for Grandad because he turned up when she was thirty and rescued her from becoming 'an old spinster'. Two of her siblings never married and they took on her family's farm and lived together for the rest of their lives. Grandad was very quiet when he first met Grandma. Everyone said he was shy, not the most romantic individual, but he offered her a respectable future. Within a few years they had three children.

Farming made money in the 1950s and 1960s, and Grandma rather liked the status of being a well-known farmer's wife. They were soon able to afford to furnish the house, which had been bare when they moved in. In the 1960s they went out to dinner dances and had a big car and spent days with friends at the horse racing. Grandma wore a mink coat and glitzy costume jewellery, and Grandad was tall and smart in his suits. I only saw the tail-end of their glory days and as I got older I could see he wasn't always kind to her. It wasn't a great love story. Everyone knew he was hard to live with. He sat in the kitchen to be waited upon by her and he drank a lot. Grandma's family and friends were her lifeline; every Tuesday when he was at the auction mart, and later in the pub, she went to get the shopping and had lunch with her friends and sisters-in-law. We called all these ladies 'aunties' as kids, regardless of their being relatives or not. Auntie Doris, Auntie Edna, Auntie Renee, Auntie Marian, Auntie Peggy. There was a kind of sisterhood between these long-suffering farm women. Grandma also regularly met her best friend, Mary Muir, who wore red lipstick and smoked cigarettes in a fancy holder. They drank brandy and played cards together. It makes me happy to look back and think she had good friendships and a little fun and glamour in her life.

Dad remembers all his cousins' birthday parties. They celebrated with games and tables laden with food. Dad was the youngest of three, and he left school at fifteen to work on the farm. He didn't have a choice in the matter as his elder brother, Norman, became sick with arthritis and went blind before he turned twenty. It was a shock to everyone. Grandma cared for Uncle Norman and he lived with them as his health deteriorated, the arthritis making him wheelchair-bound. Despite Dad running the farm for several years and doing all the work, Grandad treated him no better than the hired help. In Grandad's eyes, Dad was the cowman and he was the boss. Even in my earliest memories they didn't seem to like each other. Grandad criticised Dad if things weren't being done his way, and always looked to find fault. The soundtrack of my young life was Mum and Dad complaining about the grumpy old man behind his back.

We had a herd of black-and-white Friesian cows, and they became Dad's pride and joy. He was up at the crack of dawn every morning to milk, and out again every afternoon. He was always back in the house for supper at 6 p.m. He got up through the night to check on a cow if she was calving, and would not leave her until he knew all was well. The year had a rhythm to it: feeding sheep and cattle in the barns in the winter months, lambing and calving in the spring, silage-making in May, hay time in July and sales in the autumn. Dad kept a flock of a hundred turkeys to be ready for Christmas. From the age of about seven I helped weigh the turkeys and organise the orders. After they had been killed, plucked, hung and gutted in the barn, we laid the oven-ready birds on big sandstone tables in the cool cellar of our farmhouse. There was a flight of narrow stone steps down from the kitchen, and I always worried about falling down because it was dark and there was no handrail. For the three days before Christmas, Grandad would sit around drinking whisky with the farmers who came to collect the turkeys. The men would be oblivious to the time, and would happily take another whisky, swapping stories with Grandad. The wives would stand around in our kitchen, saying things like 'We can't stay long . . .' They would chat to Mum about how much they had to do at home, coats on, agitated and frustrated by their husbands' blathering. There were endless cups of tea, home-made mince pies, and a sink full of washing-up. There was also a Tupperware box kept in a drawer in the bureau that the cash was stuffed into. At the end of the day Grandad would open it, fold the pound notes up and tuck them into his jacket pocket. Mum would curse him as she swung the door shut, because she and Dad did all the work and he took all the money.

Outside, I liked feeding the calves best. I'd whisk up their milk-replacement powder in warm water. I loved how their pink noses snuffled and their tongues slurped it from the grey metal buckets. If they weren't used to drinking from a bucket, I'd put my fingers in and they'd suck them until they got the idea. I didn't like it when we had to dehorn them with the gas torch that Dad pressed onto their tiny buds. I sometimes held their heads but hated the burning smell.

By the time we moved to the farm, when I was three, there was only one mare left. She lived in the front field and I liked to watch her, but she was big and towered over me when I went near her, and I felt too scared to brush her. When she foaled, they were sold, as Grandad was getting too old to do the horses any more. Dad didn't like them – he said they were a hassle when there was real work to do. But when I was eight or nine I desperately wanted a pony. I went to bed every night with a book called *Show Pony* tucked under my pillow. I dreamed of riding the winning pony at a local show, all dressed up in fancy riding clothes. I spent hours reading stories of girls having adventures on horseback. But I couldn't persuade my parents to take me to riding lessons. Dad just kept saying he hated horses. It wasn't until I was much older, when I thought about it all, that I realised it might have been born from hurt and sadness, because Grandad had been so kind with the horses and so hard on him.

After an awful lot of pleading from me one year, Dad agreed to winter a pony. We borrowed her for a couple of months from Dad's cousin, who ran a local trekking centre. The pony was called Pearl. I loved brushing her and desperately tried to tame her fuzzy coat, but she always looked messy whatever I did. That winter I only rode her up and down our yard a handful of times, with Dad leading me, before he had to do the milking.

In the family album, there is one photograph of me with my grandad. I am a baby bundled in a blanket and set on his knee for a second while someone takes a photo, and he looks like he is forcing a smile. I can't remember him ever noticing me or showing me any kindness as I grew up, and I didn't feel much sadness when he died. I wasn't sure if I was supposed to cry at his funeral – no one else was. I didn't learn until much later in life that his own father had died when he was eight. It's only now, as a mother myself, that I imagine my grandad as a little boy losing his dad and know how hard that must have been for his mother. As a boy he'd picked turnips in cold fields, ploughed fields by walking behind a horse, and cut the throats of pigs, gathering the blood into pans for his mother to make black pudding. Farming boys like him had to do hard things. He had grown up fast to be a strong working man, and he took pride in the fact that he'd got a farm of his own and won shows with his horses. Family life seemed alien to him. He knew a lot about farming, but perhaps not much about love.

Until I was three, we lived in a bungalow a mile or so away from the farm. Dad came home every day for his dinner. My parents talked about the farm all the time, but Mum and I weren't actually there very much. Mum played with me and my baby brother a lot. We had tea parties and picnics and made mud pies in the flower beds. She was good fun. The garden behind the bungalow was maybe seventy feet long, but to me it seemed so big I didn't dare go to the other end. Once, Mum disturbed a sleeping fox in the garden shed. Another time she rescued a racing pigeon with a broken wing, and we fed it birdseed that we got in the pet shop. She kept the pigeon in the front porch for a week or two, but we had a black-and-white fluffy cat called Dandy and Mum thought it might kill the bird if we opened the porch door, so she would scream at anyone who went to open it without her putting the cat away first. The pigeon made a full recovery and Mum found the man it belonged to from the number on its ring.

The bungalow was Mum and Dad's first home, but also kind of not their home, because it belonged to the farm. The plan was that it would be where my grandparents retired to, so everything had to be decided by committee. Family farms often work like that – everyone in the family is in everyone else's business. When we eventually swapped houses with my grandparents it looked from the outside as if Dad was the farmer now, but Grandad was still very much in charge. He would turn up each morning and sit by the Aga in our kitchen asking how many lambs we'd had or if Dad had spread the muck or ordered the turkeys. Dad could barely eat his breakfast for answering questions. I kept out of the way when Grandad was there.

Mum's life wasn't her own any longer; managing the busy farmhouse and feeding the farm workers and us kids was a full-time job. She also had a big garden to see to. There were freshly dug potatoes and carrots to scrub, peas to shell and raspberries, rhubarb, plums and apples to deal with when they all became ripe at once. Cabbages came into the kitchen with wriggling little green caterpillars to be picked off. Mum said she'd always wanted to be a farmer's wife, but I think her dream was more like keeping a few hens and working outdoors; tending the garden was as close to that dream as she got. The farm work was all done by the men. There was no option for Mum to be anything other than a full-time housewife.

## Marmalade

I am standing on a chair at the table wearing a navy-blue corduroy pinafore with bronze buckles that my grandma sewed for me and bottle-green tights, my long T-shirt sleeves rolled up. Mum is hurriedly setting things out on the table. I turn the handle on the meat mincer as she pushes chunks of orange rind and thick white pith into the top. I try and copy her. 'Don't put your hand in there!' she scolds. The kitchen is a mess. The meat mincer is screwed to one side of the table, with a clamp with newspaper above and below to stop it marking the table. The rest of the table is covered in an old oilcloth, which can be wiped clean. Brown glazed Mason Cash bowls are full of chopped rind and the diced flesh of oranges. Muslin cloth bundles, packed full of seeds and pith and tied up with string, sit in bowls of juice. It is nearly time to start the boiling. All other work, and food, gets put on hold when we are making marmalade.

I remember that afternoon as if it were yesterday. It was a few weeks after Christmas, but the boxes of decorations were still sitting by the bottom of the stairs. Mum had been frantically cleaning all the rooms in the house, starting at the top and working her way down. Every Tuesday morning the cleaning stopped, and we'd go to the greengrocer's in town to see if the oranges had arrived. This particular morning we'd been late setting off because Dad had needed her help with a cow calving and was late in for his breakfast. At coffee time Grandma was on the phone to say, 'I've got mine, have you been for yours yet?' Flustered, Mum rushed us off into town, saying she hoped she hadn't left it too late, not wanting to be 'shown up' by her mother-in-law. We arrived at the greengrocer's and queued behind a few other determined women. Mum relaxed when she saw the stack of newly delivered boxes on the floor; she pointed towards them and said to me quietly, 'That's what we have come for.' The greengrocer's was a narrow little shop run by a husband-and-wife team who seemed to automatically know what you wanted when you walked in the door. They spent their days placing two or three pears in paper bags for old ladies, twisting the bags magically around and upside down without dropping the fruits. I liked getting a few cherries in my brown paper bag – spitting the stones back into the bag as we walked around town to get the rest of the shopping. We left the shop triumphantly with three boxes full of hard-looking oranges, all gnarly and bumpy with a leaf or two still attached. They didn't look very appealing, and they were too bitter to eat. I didn't know as a child where these oranges came from. The 'Seville Oranges' stamped on the box meant nothing to me. There were also a handful of lemons in the box and a few bags of sugar we picked up from the Co-op.

Mum didn't even take her coat off from our trip to town; she said Grandma would 'be here in a minute'. Mum was always trying to stay one step ahead of her. She went straight to the under-stair cupboard, pushing the Christmas decorations out of the way, to rummage deep amongst the jigsaw puzzles, games and old coats before hauling out a rattling box of glass jars. She hurried to the kitchen and washed all the jars with hot soapy water, while I picked last year's labels off with my fingernails. Then she set them into trays to sterilise in the oven as she worked. These glass jars were reused every year and she had her favourites; there was a variety of shapes and sizes, and, if she gave away a full jar of her precious orange-glowing treasure, people were expected to return the jar. She knew who didn't, and they weren't offered any more.

Grandma arrived with her own boxes of oranges and Mum settled into the work. Together they diced, minced, chopped and deseeded the fruit. I helped weigh out the sugar and squeeze the lemons, wincing when the juice hit my eyes from time to time. So many different things were happening at once. Grandma was in charge, instructing Mum on what to do next. I was their extra helper in this big mysterious business. I was told to put a saucer in the freezer, which baffled me. They didn't talk much; it was all about getting on. Grandma was kind, helping me learn the skills she had acquired over the years. She was an expert with a needle and thread and would show me how to embroider and make all sorts of things from fabric, often from very little. But as I grew older I saw a very different side to her: she was rather bossy and very good at getting everyone around her to do as she wanted. While she was kind to my mum in those early days of taking on her farmhouse, I came to realise that Mum had to follow an unspoken code of how to do things.

After a while, a spoonful of very hot golden liquid was scooped onto the frozen saucer and put outside on the windowsill for a minute or two to cool. And then the moment of truth . . . Grandma's finger gently scraped and pushed the blob. Her fingers were short and stubby like mine and she wore a plain gold wedding band; she had hard, working hands. The skin on the marmalade had formed: the jelly was set, it was ready. Mum scooped it from the pan with a jug and poured it through a funnel into the cooled jars. I pleaded to write on the little white sticky labels, but Mum wrote the date on them neatly. Instead, Grandma showed me how to place the little circular wax discs carefully on top of the pots, explaining that this was to seal the top from air getting to the fruit, to stop it going mouldy. Mum, like Grandma, always used thin clear plastic jam covers and elastic bands to seal the jars, rather than the gingham fabric that I admired in the shops cut into circles with pinking shears to make a crinkled edge.

Once cooled, the jars were lined up on the pantry shelves. The effort was over for another year.

I didn't really enjoy eating marmalade until I was in my twenties. Maybe because I had so much of it as a child – endless soggy, sticky marmalade sandwiches in my school packed lunches. I always saw marmalade as a lot of work and something that involved a fair bit of stress – seeing whether Mum had passed another of Grandma's farm-kitchen initiation tests. I didn't associate it with a delicious tangy conserve from posh shops until much later in my life. I didn't see making it as a process to be enjoyed and a skill to be passed on. Now, when I taste good marmalade, I am instantly back in that kitchen with those two farm women.

## GRANNY PYPER'S MARMALADE

If you like even more bits, slice up the lemon rind too, or another citrus, and mix in with the orange.

**Prep 1 hour 30 minutes**
**Cook 1 hour**

Makes 12 jars (454g size)

**Ingredients**

1.4kg/3lb Seville oranges

3 lemons

3 litres/5½ pints cold water

3kg/6lb 9oz caster or granulated sugar (570g/1.2lb to every 570ml/
   1 pint liquid)

**You will need**

A large pan, a selection of 12 glass jars with lids – or a pack of cellophane lids with wax discs and labels (found in preserving section of supermarket or online) – a muslin cloth, some string, a wooden spoon, a measuring jug, a large bowl, hand mincer or chopping board and knife, a sieve, small saucer or plate

## Method

1. Wash the glass jars in hot soapy water, rinse well and warm in a 180°C/ fan 160°C/gas 4 oven for 10 minutes to kill any bacteria. Warm them up again before you need to fill them; boiling-hot jam may break a cold jar.

2. Peel the oranges with a knife or hand peeler and halve them. Halve the lemons too but don't peel them.

3. Squeeze the juice out of the oranges and lemons and sieve it into a jug or bowl.

4. Take the pips and pulp from the sieve and put them into a muslin cloth, then tie it with string.

5. Mince through an old-fashioned hand mincer or slice and chop the pith and peel from the squeezed orange halves. Cut it coarse or fine according to your marmalade taste.

6. Put the lemon halves, orange peel and muslin bag of pips into a pan with the water along with the strained juice. You can leave this to stand overnight or just carry on to the next stage.

7. Bring the mixture to the boil and simmer for approximately 40 minutes (including the muslin bag of pips).

8. Allow to cool and remove any rougher pieces, if you want to, with a slotted spoon.

9. When cool enough, squeeze the bag of pips over the pan to release any pectin (natural thickening agent). Scrape away the pectin with a spoon or knife and stir it back into the mixture.
10. Put a saucer or small plate in the freezer.
11. Measure the mixture in a jug then return to the pan, adding 570g/1.25lb of sugar for every 570ml or pint of liquid.
12. Bring the mixture to the boil again and keep stirring for 10 minutes.
13. Put a teaspoon of the marmalade on the cold saucer to test for a set. Leave it to settle for a minute. You are looking for a wrinkly skin on it as you push it with your finger. If it doesn't wrinkle, put the plate back in the freezer and keep the marmalade boiling for another 5 minutes. Keep going until your marmalade wrinkles on the plate.
14. Pour the marmalade into warm sterilised jars and top with a wax disc to seal, and either a fabric cover or cellophane one – these are usually included in the pack with the wax discs and elastic bands – or a jam jar lid.
15. Label the jars with the date. They will keep for over a year if properly sterilised and sealed and kept in a cool dry place.

Grandma was always in and out of our kitchen when I was little. It had of course been, until recently, her kitchen. But she and Mum didn't work alongside each other, except for making marmalade and the Christmas cake each year. I didn't know at the time that Mum had never done any of this kind of work before. She had arrived in this world where all the women seemed to know all of the things that women like them were meant to know, yet somehow she never seemed intimidated. She had more or less no cooking skills, because her mother had taught her nothing. Through a mixture of fast learning and winging it, she had got through those early years on the farm. I would often hear her mutter under her breath when Grandma overstepped the mark with well-meaning advice. By my teenage years, Mum had learned how to bake a better Christmas cake than Grandma, and everyone loved her marmalade. Every

January she would often work late into the night, and when Grandma asked her 'Have you got your oranges yet?' Mum would appear with a couple of jars made the week before and casually say, 'Yes, would you like some?'

My dad never cooked anything – apart from a pan of custard because he said he could make it better than Mum. He learned to make that on Tuesdays because that was the day my grandma left a cold lunch for the men and went to do the shopping and meet her friends in town. Grandad would be in the Agricultural Hotel boozing and she would drive him home. On Tuesdays it was a plate of bread, cold ham or beef, or cheese, and a tomato cut in half, all seasoned heavily with salt and pepper. They always wanted a piece of cake or tinned fruit with custard afterwards to make it into a 'proper dinner'.

## CUSTARD

### HOME-MADE CUSTARD

**Prep 10 minutes**
**Cook 25 minutes**

Makes approx. 800ml/1½ pints

**Ingredients**

570ml/1 pint whole milk
200ml/7fl oz double cream
1 tsp vanilla extract or a vanilla pod
4 large egg yolks
3 tbsp cornflour
100g/4oz caster sugar

**Method**

1. Pour the milk and cream into a pan and add the vanilla. If using a vanilla pod, split the pod with a sharp knife and scrape the seeds into the milk/cream mix along with the pod.

2. Warm the milk/cream over a medium heat. Bring to a simmer but do not boil.
3. In a bowl, whisk the egg yolks, cornflour and sugar together.
4. As the milk/cream is nearly boiling, pull it off the heat for 2 minutes. Carefully remove the vanilla pod (if used); this can be washed and dried, and put into a jar of sugar to make vanilla sugar.
5. Pour the milk/cream mix over the egg, cornflour and sugar paste. Whisk together quickly.
6. Pour into a clean pan and gently warm, stirring for about 20 minutes until thickened. Serve immediately or drape cling film over the surface of the custard to prevent a skin forming, allow to cool then keep covered in the fridge for up to three days.

## CUSTARD MADE WITH BIRD'S CUSTARD POWDER

Makes approx. 570ml/1 pint

**Ingredients**

570ml/1 pint whole milk

2 heaped tbsp custard powder

1 tbsp sugar

**Method**

1. Pour the milk into a pan.
2. In a bowl or jug blend the custard powder and sugar with 2–3 tbsp of the milk from the pan to form a paste.
3. Bring the milk to the boil until it is rising in the pan, then lift off to pour into the paste in the bowl or jug, stirring it fast. It should thicken quickly.
4. Leave to stand for 2–3 minutes before serving.

## MACAROONS (TO USE UP YOUR EGG WHITES)

**Prep 20 minutes**
**Cook 10 minutes**

Makes 24

Ingredients

    2 egg whites

    125g/4oz ground almonds

    175g/6oz caster sugar

    ½ tsp almond (or vanilla) essence

Method

1. Heat the oven to 170°C/fan 150°C/gas 3.
2. Whisk the egg whites in a clean, dry bowl until they have formed stiff peaks.
3. Fold in the ground almonds, sugar and essence with a spoon until combined.
4. Place evenly spaced 12g blobs of the mixture onto a greaseproof-lined baking sheet and bake in the oven for 10 minutes or until lightly browned and firm to the touch. The crusts will harden as they cool.
5. Leave to cool and store in an airtight tin.

Mealtimes anchored the days. With very little experience, Mum had to find her way around managing this big farmhouse and doing all the jobs, mostly by herself. There was no money to pay a maid to live in the attic, like there had been in Grandma's day. Life was changing. Mum soon refused to put a cooked meal on the table at twelve noon every day for the men. 'Enough is enough,' I heard her say. She told the farm workers to bring packed lunches and told my grandad to have his at home. Dad got a cheese sandwich, and custard (if he made it himself). Her next blow was to inform my grandad he couldn't just come into 'her' kitchen and sit by the Aga for as long as he liked. I was glad about this, because I didn't like him being in our house. A few years later, Grandma only came on Fridays, when she came to get her hair done by Mum. As Grandma leaned over the kitchen sink with a towel around her shoulders, Mum would pour jugs of water over her head from a basin and rub shampoo into her grey curly hair. Grandma always looked much older to me when she turned around with wet hair.

If Mum was irritated with her, the water would accidentally be too hot or too cold. Mum then combed and pinned her hair in rollers and blow-dried it under a hair-drying machine that sat on the edge of the kitchen table and was placed over her head like a space helmet. Grandma would sit under the dryer, often scowling as she watched Mum work. Because Grandma was deafened by the blower, Mum would call her all the names under the sun, her response to Grandma's endless criticisms and comments. But Grandma brought fresh baking with her, and we were always eager to see what was in her tins and tubs – apple pasties, ginger biscuits, madeleines and gingerbread. Her coconut and jam tarts were my favourite. Mum reckoned this baking was a thinly veiled insult, the unspoken insinuation being that her precious son might waste away in this renegade house otherwise.

## COCONUT AND JAM TARTS

Prep 20 minutes

Cook 20–25 minutes

Makes 24 small tarts (you will need to make 2 batches unless you have 2 cupcake tins)

Ingredients

For the pastry:

400g/14oz plain flour, plus extra for dusting

½ tsp salt (unless using salted butter)

200g/7oz cold unsalted butter (or 4oz lard + 3oz butter)

1½ tsp caster sugar

5–8 tbsp cold water

Of course, use ready-made pastry if you want to, but the joy of making your own is knowing that it doesn't contain a lot of hidden ingredients.

For the filling:

    2 eggs

    125g/4oz caster sugar

    50g/2oz melted butter

    1 tsp vanilla extract

    225g/8oz desiccated coconut

    175g/6oz raspberry jam (for a quick home-made recipe, see p.164)

Method

1. Make the pastry. If you have a food processor, whizz up the flour, salt, butter and sugar until it resembles fine breadcrumbs. Otherwise, rub the butter into the flour, salt and sugar lightly with your fingertips in a bowl.

2. Adding 1 tbsp of water at a time, pulse the mixture until combined into a ball, or stir in with a fork until you are starting to form a ball of pastry.

3. Tip the pastry out of the mixer or the bowl onto a clean, lightly floured surface, and combine the rest by hand.

4. Once in a ball, flatten into a disc, cover, and chill for 30 minutes.

5. Heat the oven to 190°C/fan 170°C/gas 5.

6. To make the filling, beat the eggs well with a whisk. Add the sugar, melted butter, vanilla and coconut to the beaten eggs and mix to combine.

7. Grease and flour the cupcake tins.

8. Roll the pastry to ½cm thick or as thin as possible, cut into 24 circles of 8cm and gently press the circles of pastry in.

9. Put a scant teaspoon of raspberry jam in the bottom of each pastry case.

10. Cover the jam with 2 tbsp of the coconut filling. Lightly press the mixture down if you're after neater tarts or leave them with a bit of texture.

11. Bake for about 20–25 minutes, until lightly golden brown.

12. Leave to cool before removing from the tin (the residual heat from the tin helps to crisp up the base of the pastry shells).

The whole farmhouse was freezing cold, except in the kitchen by the Aga. When I went to bed I huddled up in my blankets, often sitting on my pillow, scared of going to sleep because of my recurring nightmare about hens at the bottom of the bed pecking my toes – even though we didn't have any hens. Ice formed in pretty patterns on the inside of my bedroom window in winter. I started my period aged eleven, and for the first few months I'd go downstairs in the middle of the night with awful tummy cramps. I'd sit with my back to the warm oven, quietly crying in pain and holding a hot-water bottle to try and get some comfort.

Under Mum's instructions, the outdated farmhouse got a make-over. She wanted to earn her own money, separate from the farm, so she and Dad decorated two rooms for bed and breakfast, and had an en suite shower room installed in the old dressing room of one of the front bedrooms. We were shunted out of the good bits of the house to make room for all this. I had to wash my hair in what we called the 'back kitchen', a room near the back door where smelly boots and damp coats were thrown off before the men entered the warmth of the main kitchen. A large chest freezer full of home-produced meat sat under the window, and there was a coalhouse in the corner.

Nudging empty cans of dog meat and thin leftover bars of dried Fairy soap out of the way, I stand on a wobbly wooden stool and dip my head over the grotty metal sink. The taps are grimy from the men's hands. I push the two ends of the rubber shower hose onto the hot and cold taps, securing the ends as tight as I can so that, when I turn the taps on, they don't fall off. Getting the temperature right is tricky and I am not allowed to use a lot of hot water, Mum always reminds me. The basic plastic showerhead barely gets all my shoulder-length hair wet properly but I lather it quickly with a two-in-one shampoo and rush-rinse it before someone draws the hot water from another place in the house and makes my water go cold.

Upstairs, the 'main bathroom' is immaculate, but we are forbidden to go in. The white porcelain bath and sink are polished dry with a towel, the silver taps are shiny. A new bar of soap sits freshly unwrapped on the basin. The toilet paper is folded into a V. Mum has vacuumed the carpet in neat lines like a football pitch on TV, and fluffy towels hang on the warm rail.

Through the B&B season, our family mealtimes became chaotic. Mum was always up at the crack of dawn, fretting over the presentation of her cooked breakfasts for the guests. She often stripped the beds before the last visitors had driven to the bottom of our lane and always cleaned the rooms thoroughly. She cursed the phone ringing with potential bookings when she was upstairs making the beds and had to run down and answer it. She set out fresh milk in jugs and biscuits on tea trays in the bedrooms before guests arrived, and spent hours washing, drying and ironing sheets, duvet covers and pillowcases. She would chat politely for hours to old holidaying couples who had nothing better to do. She also spent a lot of time on more eccentric issues, like which teaspoons to put in little bowls of jam and marmalade. She vented at us when she was exhausted and it was all getting too much. By late afternoon, she no longer much cared what we were having for supper.

Once, as I retreated through the house on tiptoe after washing my hair, I heard Mum at the front door saying, in her poshest voice, 'Oh, I'm sorry, we don't have any vacancies tonight, but I can give my friend a call and see if she has a room.' No vacancies? Both of the rooms were empty. I wanted to run back downstairs and shout 'Yes we do!' but I kept out the way. Mum had her own special system of running a B&B, and it didn't involve the rest of us. I asked her, later, in front of my dad, 'Why didn't you take them in?' She snapped at me that the pillowcases were not ironed yet, the room hadn't been hoovered, and she didn't have enough mushrooms in the fridge for breakfast. None of this made any sense to me because the rooms were fine. Dad said, 'We need the bloody money', but Mum had her own rules. She never did any paperwork, paid any bills or had a clue about the farm finances. Dad did all of that by emptying out an overflowing drawer from the bureau onto the kitchen table once a month and writing cheques until late at night.

As a teenager I spent most of my time in my room in the attic. Mum hardly ever came up the three flights of stairs. She had enough on her plate focusing on the bits of the house that other people would see. There were bare rooms alongside mine in the eaves that had once been bedrooms for the staff of the farm, in my grandparents' time. There had been two farm lads and a maid called Nelly. All the meals in the farmhouse had been made from scratch, so the growing, harvesting, preparation, preserving and storing, along with the cooking, had been relentless – as was the washing and cleaning without modern machines. Grandma had needed Nelly's help to do it all. They fed the family of five plus two farm men, and a constant flow of visitors, particularly at lambing, clipping and hay-making time. When Nelly left, Grandma took on a young woman called Joyce. She came a few days a week to work in the house and spent a lot of time looking after my dad when he was little. Now all these workers were little more than ghosts, but they'd left the attic for me. I often wondered about girls who'd lived in attics in houses like this, and whether they'd been treated well by the farm women like my grandma. Their stories didn't ever get told. The only stories we heard centred upon the big men like my grandad, or on the horses or cattle that won rosettes at the sales and shows.

As I got older, I was expected to do more jobs to help Mum. After school I'd get kindling sticks from the woodshed to light the fire in the sitting room. I set and cleared the table at suppertime. I had to tidy the worktops, wash up and take things back to the pantry. And when Mum took the washing down from the rack above the Aga and folded it, I carried it upstairs. My brother was never asked to help in the house. This infuriated me, and I always protested about him getting away with doing nothing. He didn't work outside on the farm at all, like boys had once been expected to, so I didn't understand why he wasn't asked to put the shopping away, or ever wash up. He seemed to be allowed to do whatever he liked. One day, when I was asked after school to go and get the dryer for Grandma out of the cupboard in the bathroom, I snapped and said, 'No.' It turned into a horrible shouty row – Mum raged at me for showing no respect to her in front of Grandma, and I was sent to my room for the rest of the evening.

I resented having guests in the house. We had to tiptoe around and not slam doors or play any loud music. Also, if Mum was supposedly running it as a business, I couldn't understand why she didn't try to make more money. The lack of focus on this drove me insane. Everything she did seemed haphazard and disconnected; she got lost in tiny presentation details that just didn't matter. To me it was simple: once we were open, from Easter to September, every bed, every night, should be filled.

### Carbonara (*of sorts*)

It is 5 p.m., long after school has finished. We are all hungry, and Mum is nowhere to be found. *Ready Steady Cook* is on the TV on the kitchen shelf. I'm glued to the little screen, watching the celebrity chefs race around behind the counter griddling steaks or fish, tossing beans, cherry tomatoes and shallots in a vinaigrette they just made. Then they crush meringues, whip cream and fold in raspberries in seconds for a simple dessert.

I can't wait any longer. I take my wooden chair into the pantry to stand on and look through the cupboards. The room is bitterly cold, so cold that I can see my breath. I look around and see food everywhere, but nothing I want to eat. I need something to make a meal with. The cupboards stretch from floor to ceiling and are glossy from many layers of thick green paint. I am faced with the sight of hundreds of tins of Princes pear halves in syrup. It looks like my mother has stockpiled tinned fruit for the apocalypse, but really it's the spoils from an accident. When a lorry spilled its load onto the road near our farm, my grandparents rushed to load up a trailer with the damaged goods, saying, 'It'll only get thrown away.' We have been slowly munching our way through the dented tins of overly sweet fruit for the last three years and we still have a way to go.

In amongst the pears, I see packets of flour, sugar, oats and cheese crackers. Several flavours of blancmange powder, a flan case, three tins of baked beans. Tinfoil trays and Tupperware tubs tumble out onto the floor as I rummage. Nothing yet. On the left of the door is a large, fixed stone table, a dumping ground for leftovers: a bowl of cooked mince, half a stale sponge cake, empty biscuit tins and a plate of last night's supper that should have been fed to the dog. I try the fridge. Here the shelves hold a large catering pack of bacon, lumps of badly wrapped cheese and a few nearly empty packets of sliced bread. More bowls of leftovers – mashed potatoes, broth and salad at various stages of decay. The vegetable drawer contains half a slimy cucumber, a few limp carrots, a head of broccoli and lots of mushrooms and tomatoes that I am not allowed to touch because they are 'for the guests'.

SEE P.293 FOR **PANTRY STAPLES**

I am still hunting for food and inspiration when I hear Dad coming into the kitchen from milking the cows. 'Where's your mum?' he asks my brother, who has now taken over the TV and hooked his Sega Mega Drive up to it. No answer. I hear the tap running and the familiar sound of the kettle boiling as Dad makes himself a coffee. I take the milk through with a few things I have found. My little sister pauses her play with her Cabbage Patch doll and looks up to me. She says, 'I'm hungry.'

I take a head of broccoli, some bacon, a few forbidden mushrooms, cream and pasta, assemble my ingredients on the kitchen table and get to work. Dad reads the paper and my brother sits playing *Sonic the Hedgehog* on the tiny TV screen. I feel like a bit of an alien in my family; I'm not sure any of them actually notice me.

I put the broccoli and pasta in a saucepan with cold water and it takes a long time to boil. I sizzle the bacon – cutting it up with scissors like they do on the TV – add the mushrooms to the frying pan and a teaspoon of mustard, and swirl in the cream. It bubbles up, turning from silky white to a sort of muddy colour. I drain the pasta and broccoli and mix the whole lot together in the frying pan, then serve it with a few bits of toasted bread crusts. This isn't just food, this is my food, and cooking it gives me independence, a sense of being free from under my mother's control. This plain little dish of pasta and sauce is an act of rebellion, and I love making it.

My brother Stuart glances at the offering and opts for some toast and cheese. My sister Alison has the pasta but starts to pick out the broccoli. As we are nearly finished, Mum appears from the garden with dirty gloves and a bucket of weeds. 'Oh, it's suppertime – I lost track of time.' She scoops the remainder of the mixture from the pan onto her cold plate and tucks in.

I took on the role of regularly cooking our family meal from then onwards. I saw myself as the one with the rational mind about how to stock the pantry and found myself regularly cleaning it out, reorganising the shelves and adding things to Mum's shopping lists. I had watched lots of cooking shows to feel confident creating meals from some kind of meat, a vegetable and pasta/rice or potatoes. I was so self-assured. I was going to show my mum how easy it could all be. I was tired of her food. I didn't want to eat grey lumpy mashed potato four times a week. Or a dried-out lamb chop. Or stringy braising steak thick with lumps of flour and dark-brown gravy and coagulated fat. Or, worse still, the latest offering from the frozen aisle: Findus Crispy Pancakes with rubbery bits of ham and a salty cheesy goo in the centre.

I didn't want to go looking for her after school in order to ask 'What's for supper?' I had no idea why she didn't know how to make food taste better, why she didn't want to be in the kitchen cooking, using fresh herbs from the garden, and why she didn't want to be with us.

As a child I had no concept of how much work she had to do and why being in the garden was probably her escape from all things domestic, her way of surviving it all. I know now how happy gardening makes her.

SEE P.295 FOR MEALS WE SHOULD NEVER FEEL GUILTY ABOUT

I wanted to take charge of the kitchen. I wanted to eat pan-fried chicken with couscous and lemon, or honey-glazed pork chops and crispy roasted potatoes. My cooking didn't always turn out like the food on the TV shows, but some of it was delicious. And there was a thrill in it when my scones rose pleasingly and everyone enjoyed them with butter and jam, or when I made a roast dinner for the first time and the beef was still pink in the middle, not dry and chewy. And perhaps there was also a teenage, slightly mean thrill in doing something better than my mum did.

I got a job in a local cafe when I was fourteen. It was here that I learned how a lot of kitchen work could get done by streamlining certain things, like making big trays of cakes, or big batches of soups and casseroles and freezing them. I saw how food could look and taste better. But Mum didn't want my ideas. She only wanted my help to clean up and carry the breakfasts through to the 'dining room'. To my cruel teenage eyes, it all seemed a scam, pretending things were perfect for guests when it was such a shambles behind the scenes. Mum was always 'popping into town' for bits of shopping, sometimes more than once a day. She was always late to pick us up from school. She never seemed to plan a meal or do a big shop to stock up properly. In the mornings when we had guests staying, she watched the clock like a hawk, and if she couldn't hear footsteps on the stairs or the creak of the dining-room door at 8 a.m., I was sent up to give the guests a knock on their bedroom door and say loudly, in my polite voice, 'Breakfast is ready.' The guests were served dry-cured bacon and free-range eggs while we would be helping ourselves to Pop-Tarts and Coco Pops.

Mum wasn't particularly secretive about her past, she just didn't talk about it much. Come to think of it, we never talked about anything openly in our family. Between Mum and me there was either shouting or silence. She'd get in a panic about guests arriving and send me to hoover an already immaculate room, and I'd say, 'Don't be stupid, it doesn't need to be hoovered.' She'd be furious and yell at me up the stairs to do it, and I'd yell back that I wouldn't. And then came the silence between us. I just learned to do my own thing, but slowly I began to wonder about my mum. I didn't know why we didn't go to see her mum, my gran, in Scotland, so I began to fumble my way towards understanding her previous life.

The salty wind hits my face as I walk down the metal steps and onto the ramp off the ferry. Cars flood past me, their headlights all illuminating the sign 'Toutes Directions'. I pull my hood up and hang back a little to take in my surroundings. I am sixteen and have just arrived in Brittany, where I am going to stay with a French pen pal called Sandie. I have my giant rucksack tightly strapped to my back; if I tip too far forward, I'll land on the ground and look like a tortoise. I grip the small bag with my money in front of me and tuck my scarf up to hide my face. I follow the steady flow of the few foot passengers, as they all seem to know where they are going, and then I see the car park ahead. As I get closer, a man and a girl with shiny dark hair blowing in the wind wave towards me, smiling, and I breathe a huge sigh of relief. I have made it and they are here for me.

Just after Christmas our teachers stressed the importance of finding a good placement for our 'work experience' week in April. Last year I had chosen to work at our local newspaper, fancying myself as a journalist, but it soon turned into the most tedious week of my life. I was given the task of writing up the 'Births, Marriages and Deaths' section, which was kind of funny as my grandma and her friends were eager readers and she would earnestly show me announcements from these pages – 'Mrs Strong has died', as if I knew who that was. A very boring man instructed me in how to type up the information sent to the paper for that week. People paid per letter, so it all had to be as they wanted, but in the house style. I was given a few examples and told not to stray into using anything other than their particular format – facts, dates, and everything in the right order.

This took me all of two hours on the first morning, and then I had nothing to do. I hung around, photocopied a few pages of something and then went to sit in the local court with a reporter to hear the verdict on a drunk-and-disorderly offence. The rest of the week was exactly the same. Every lunchtime I went and sat in the town library for the hour, sneaking a sandwich in and reading a romance novel by the radiator.

This year I was determined I would do something more ambitious, more adventurous, more exciting. When I came up with my plan in February, I told Mum and Dad. I have no memory of how they reacted, perhaps because I wasn't remotely interested in their opinion. I told them I was going to France for two weeks to teach English in a French school.

They didn't stop me.

*

The journey starts with an eight-hour train ride from my home in Penrith to Plymouth, changing at Birmingham. To catch the early-morning ferry tomorrow I will need to stay overnight in Plymouth, near the ferry terminal. I find a B&B overlooking the harbour by phoning the tourist information office in Plymouth.

I get off the train and look around. I have never been this far away from home on my own before. I look at my map: the B&B is only two miles through town and along a main road. I can do this. I strap my rucksack tightly on my back and set off walking, hoping to arrive before it gets too dark.

My heart is pounding by the time I reach the dual carriageway. I want to walk faster. Cars race by me as I stop and figure out which direction to go in. I follow a hunch and find an underpass that smells of urine to cross to the other side. I keep my head down as I walk, hoping there is no one lurking in the shadows. I am relieved to get back into the fresh air and cross to where the streetlamps are.

I am hungry but all the shops and cafes are well behind me now. I find a stray Polo mint in the bottom of my coat pocket and suck it for reassurance. I stop to study my little tourist information map in the fading light; I seem to have ended up at the furthest end of the street that I need to be on, so I take a deep breath, knowing there is nothing else to do but walk. The buildings are big cream-painted villas with bay windows and curtains drawn. I finally see a sign, with the name of the place I am looking for, swinging in the wind. I pause, walk up the steps and boldly ring the bell. A man with a gold chain and a shirt unbuttoned too far opens the front door. I squeeze past him to get inside; his breath smells of cigarette smoke.

He asks me to sign in the book with my name and address. He tells me the bathroom is across the hallway from my room on the second floor, and that breakfast is at 8 a.m.

I relax when I see his wife pop her head through the door down the hallway. She says hello in a soft welcoming tone. She asks how far I have come. I say the Lake District. 'On your own, dear?' she replies in a surprised voice.

I nod and say I have to catch a ferry to France at 7.30 a.m. so won't need breakfast. The man says he will book me a taxi to be waiting outside at 6.45 a.m. and wishes me a good trip. I suddenly feel very young and a bit scared.

My room, number 5, is decorated with floral wallpaper and has a single bed with a pink candlewick bedspread in the middle of the room. There is a basin and mirror in the corner, and on top of the chest of drawers a tea tray set out like Mum does at home for the guests. There is a kettle, a cup, some UHT milk and a couple of mini-packs of biscuits, which I eat immediately. I take off my rucksack, coat and boots and sit on the side of the bed, not wanting to mess it up. I look up at the shiny brass light fitting and wonder why I thought this was a good idea. I wish I had asked the man if I could use his telephone to call home, but it's probably best that I don't hear Mum's voice tonight.

I hardly unpack anything, only my toothbrush and pyjamas, leaving my clothes for the morning set out on the only chair in the room. I check my ferry ticket and zip it neatly folded back into my coat pocket. I climb into bed and the bedspread tries to slide off if I move around, so I lie still and watch the glow of the streetlamp through the gap in the curtains. The sound of a cat meowing nearby keeps me awake. I hear doors opening and closing and toilets being flushed, showers being run. I listen for footsteps near my door. I am frightened that I will miss the ferry if I fall asleep.

I check my watch repeatedly through the night and when it turns 6.20 a.m. I start to get dressed. I pack up and shut my door quietly, as the rest of the house is sleeping. A bloke pulls his car up to the B&B and says out the window, 'The ferry, luv?' in a friendly accent. I smile and nod. I climb into the back of his Ford Sierra with my bag and he speeds along the empty roads, passing only a milk float. I pay him a couple of pounds and wave as he drives off, as if he is family, then walk across to join the small queue waiting to board. It takes ten minutes for my ticket to be checked and then I am on the boat. First stop: breakfast. I haven't eaten since my journey yesterday, when I munched on Mum's ham sandwich and the piece of fruitcake that she gave me for the train. I am ravenous.

I take a tray and load it up with a bowl, cereal, milk and a banana, order a whole cooked breakfast with orange juice, toast and coffee, and then go back for one of the fancy pastries that has a sign reading 'Pain au Chocolat'. I wrap this rare treat up in a napkin for later.

I walk around the ferry looking for a comfortable chair in a quiet spot before settling on a lounge with a few old folk snoozing. I tuck my rucksack under my feet, lean back into the headrest, watch the vast grey sea out of the window and nod off to sleep. It is 3 p.m. when I wake to people bustling around me, getting their belongings. I nip to the loo before we queue to get off the ferry, and splash water on my face to freshen up.

Sandie's house is only about a twenty-minute drive away from the ferry terminal of Roscoff in Brittany. It is warm and welcoming when we go in. Her mum doesn't speak English but she hugs me when we get inside. She shows me the bathroom and has made up the spare bed in Sandie's room nicely for me, with white sheets and a lovely rose-patterned quilt. We have been pen pals for just over a year and my French is good enough for basic conversation. Sandie's mum seems impressed and asks about my family, and I give them gifts from home, toffee and fudge and a Lake District tea towel with sheep on it.

I ask to use the telephone to let everyone at home know I am safe. 'I'm here,' I say, 'Yes, the ferry was OK', and Mum tells me the latest news about the lambing and what Grandma said to her that wound her up when she came for her regular Friday hairdo. I cut her off by saying, 'It's probably costing a lot, this phone call – I'll ring you at the weekend. Bye.'

At 6.30 every morning we dip baguettes into deep French bowls of hot chocolate made with milk and cocoa in a pan. Sandie's mum cleans up the kitchen before she goes to work in the local bank. Her dad drops us off at school before he goes to work in blue overalls. We are all out the door by 7.30. The French high school is modern with large windows. All the kids stand around smoking before they go into class; I decline their offers of a cigarette. I listen to their gossip, trying to figure out what they are saying. I get a few curious looks from the boys. When I take my place at the back of the classrooms, the teachers kindly give me sheets of what they are studying, but none of it really makes any sense. I take part in the English lessons by reading aloud to the students and helping them spell their words correctly or form better sentences.

## COCOA AND HOT CHOCOLATE

Hot chocolate and hot cocoa are two different drinks. Hot chocolate in its simplest form is melted chocolate in hot milk – creamy and delicious. But if you want a lighter drink, cocoa can be sweetened to your taste with sugar and stirred into hot milk. Both drinks can be flavoured too with a variety of things like orange, mint, caramel, chilli or spices like cinnamon and nutmeg.

### LUXURY HOT CHOCOLATE

**Ingredients**

100g/4oz good-quality chocolate (semi-sweet, 30–40% cacao; darker chocolate will give a richer, bitter flavour)
1 large mug of whole milk

**Method**

1. Chop or grate a small bar of chocolate into a pan of hot milk until it is all melted, then pour into a mug.
2. Top with whipped cream and grated chocolate.

COCOA

Great for kids as it gets them a warm chocolate drink without all the sugar of the shop-bought powders.

### Ingredients

1 mug of whole milk

2 tbsp unsweetened cocoa powder

½ tsp sugar (more or less, according to taste)

a pinch of salt (again, to taste)

### Method

1. Pour the milk into a small pan and set over a medium heat.
2. Whisk in the cocoa powder and add the sugar and salt, if using, and any flavourings.
3. Heat to dissolve the sugar but don't boil.
4. Serve in a mug with a sprinkle of cocoa.

Lunchtime is the best part of the day. We take trays as we enter the huge canteen and Sandie shows me that I have to take a cold salad first, then go to the hot counter to get a plateful of steaming beef, pork or roasted chicken dripping with delicious sauces, and potatoes in cream or roasted with herbs; there is couscous or tagliatelle and haricots verts and baby carrots. The desserts are laid out on another chilled counter: tarte aux pommes or chocolate delice and crème caramel. I am in heaven. I have no idea what to choose because it all looks so good. Over the two weeks, I work my way through the menu. The lunch break is long. No one has a packed lunch and there are no silver trays of greasy chips, pieces of chicken in breadcrumbs or dodgy-looking hot dogs or frozen burgers with rubbery cheese.

After sport in the afternoons and a bit of hanging out at the park with Sandie's friends, we go back to the house. We munch on baguettes with Nutella and salty crisps. Sandie's mum cooks every night. Her food is like eating in a restaurant but served in a very simple way. We tuck into medium-rare steaks sliced thinly and stirred into pasta with a side of green salad in a tasty vinaigrette; or breaded escalopes of veal with frites; cassoulet; and some kind of fish, steamed with leeks and pommes Anna – sliced potatoes layered up and cooked in butter.

## DAUPHINOISE POTATOES

This is an easy dish to make in advance of your meal. You can either keep it warm or put it together, leave to cool, store in the fridge and bake when you need it. It makes a change from mashed or roasted potatoes and goes well with most roast meats. We have dauphinoise potatoes with roast lamb, lamb leg steaks, ham or shredded ham hock, rump or sirloin steak and roast pork.

**Prep 20 minutes**

**Cook 55 minutes**

Serves 6–8

### Ingredients

570ml/1 pint double cream

570ml/1 pint whole milk

3 garlic cloves, peeled and bashed whole

8–10 large potatoes (Maris Piper or King Edward floury potatoes)

salt and pepper, to taste

a handful of grated cheese, roughly 50g (optional)

### Method

1. Heat the oven to 180°C/fan 160°C/gas 4.
2. Put the cream, milk and garlic cloves in a large heavy-bottomed pan on a medium heat to warm through.
3. Peel and slice the potatoes into 3–4mm slices.
4. Put the potatoes into the pan of warm milk and cream and heat until simmering for 5 minutes, stirring occasionally to stop them sticking together. Season well with salt and pepper.

5. With a slotted spoon, gently scoop the potatoes into a large, buttered baking dish/tray (roughly 2 litres in volume) and carefully pour over the remaining milk/cream. Sprinkle over the cheese, if using.
6. Place the tray in the oven to bake for 40–45 minutes or until golden brown.

At the weekend Sandie's mum drives us into town for the Saturday market, and we help carry shopping bags laden with leafy vegetables, eggs, fish, meat, and cheese from the fromagerie. We buy crunchy baguettes jambon-beurre for lunch and warm pains aux raisins. I came to France to teach English, but now here I am much more interested in the food.

When I get back home, my friends can't believe I went all that way on my own. I don't let on how scared I felt at times. Mum and Dad have had blackleg among the bullocks, and several have died. They are so stressed that they barely notice I am back. I still live in their house, but I kind of feel like a stranger. Their world isn't mine: I have been away, seen another life, and I liked it.

I start my A-levels later that year and buy a maroon-coloured wool blazer that I see on a model in the Benetton shop window.

My French trip has given me a confidence that I didn't have before. I start to embrace being a bit different. I never thought of myself as pretty. At primary school I was teased mercilessly about my 'ginger' hair. I had freckles all over my face and arms and my ears stuck out so far that, when I complained about them, Mum suggested I could get an operation to have them pinned back. My milk-bottle-white legs and arms shone out in summertime amongst the tanned skin of my friends. If I ever wore a swimming costume at an outdoor pool, the other kids would run around soaking up the sun and I just went red and burned. It made me hate summer; it made me hate myself.

As I get older, I can see the fakeness in the tanning creams and perms that all the girls experiment with. The desire to look and be like everyone else drifts away from me. I buy my baguette from the new sandwich shop in town, meet up with a friend and we hang out at her boyfriend's flat.

## Bacon Butties

It is 7 p.m. on a rainy Thursday night. My friend Helen phones to tell me she wants to go out and I have to be ready in twenty minutes. She says, 'It's a leaving do for two young farmers, they're going to Australia – come on, it'll be fun.' She knows I'm not that keen, but I don't tell her I'd rather stay in and read. My boyfriend has gone away to university and Helen thinks I'm lonely. She has just passed her driving test and will find any excuse to go out. 'OK,' I say, 'see you soon.'

I have a part-time job waitressing and making bacon butties in the local auction mart cafe, and, while I like the people I work with and the job, I don't like being in amongst lots of young farmers. The boys mostly act like idiots, daring each other to see who can drink the most and then doing silly stunts. The girls act like this is all very funny, also drink a lot, and try to get off with the coolest lads. It's a tribe and I don't consider myself part of it. Most of them think I'm weird and snooty anyway, marked out by my maroon blazer, A-level aspirations and unspoken, but obvious, desire to leave this town. They know I'm not interested in tractors or stories of who shagged who last weekend, so I'm a bit of an outsider despite living on a farm like they all do. I don't care who the leaving do is for – I just tag along to keep my friend company.

SEE P.294 FOR BACON

I brush my hair, dab my eyes with some shiny eyeshadow and think that, even though I'm not that bothered about going, I'll make an effort; it's a good excuse to wear my new black turtleneck top with my high-waisted jeans. Helen picks me up and drives us to a pub on the edge of town in her pale-blue mini Metro.

When we get inside it is really busy. I recognise a few faces. I squeeze to the bar and buy Helen a Coke, and half a cider for myself. She chats to so-and-so and we stand around, me trying to look happy to be here. Then I notice the older brother of a girl I work with. He's standing at the bar and everyone around him is laughing at something he has said. Another friend pats him on the back and passes him a drink, and I realise he is one of the two lads going away. I've seen him when I've stayed over at his sister's house, and now and again at the auction. He didn't really notice me the first couple of times I went – he was mostly busy working with his dad or sitting splayed across a chair in the kitchen, reading a book, which for the farm lads I knew was kind of unusual. But the last time I went over, he joined in with our Scrabble game. His sister hated being beaten so went to bed before us, and we carried on playing until well after midnight. And now he is at the bar, and he glances over to us and I can't look away. He has straw-blond hair and gorgeous blue eyes. He smiles at me with a nod of hello and I go over to wish him well on his trip. I ask where they are flying from tomorrow and laugh when he tells me that he hasn't packed yet. He buys me a drink and suggests we sit at a table in the corner. Helen glances over at us, grinning, and I try to ignore her. We end up chatting for the rest of the evening. He isn't like the rest of the lads. I sense he feels as awkward as I do at this kind of thing, but is better at hiding it by playing the fool. The evening passes quickly. I have no recollection of anyone else being in that pub. Nor do I remember going home. I only know one thing for sure: it's going to be a long time until he's back next February.

We didn't kiss that evening, but I lie awake wishing we had, hoping he's thinking about me too.

I turn eighteen a couple of weeks later, in December, and I finish with my long-term boyfriend. I carry on doing my own thing at school. I dodge all PE lessons. I hate wearing the really short school PE skirts and pants, showing my bare white legs. I'm never able to keep up with the other girls at running. Hockey is terrifying. I spend all my lunch breaks in the art department doing my coursework.

Then, just after Christmas, I get a postcard from Adelaide. Mum and Dad read it before I can. 'Why is *he* sending you a postcard?' they want to know as I race to pull it off them. They recognise his name, know his parents and tell me that our grandmothers are good friends. He wrote it from the stands at the Grand Prix in Adelaide. 'Vroom . . . vroom,' he wrote, describing the noise and the petrolheads around him. He wrote that he'd been on Bondi Beach with surfer dudes, eating a ham sandwich for his Christmas dinner, and signed it: 'Love James'.

I count the days until he is back.

I am wearing the cutest dress I have, a silver, swishy, low-cut number that makes me feel great, with black tights, boots and my neat denim jacket. I am out with a few friends in town. I heard he was going out tonight from his sister in the sixth-form common room. The pub is noisy and crowded. I lean in to kiss his cheek to say hi. He doesn't lean away, and I look into his eyes and feel a spark between us. One of the lads makes a silly whistle in the background. They are all being stupid, so I blush and quickly get a drink and catch up with a couple of other people I know. He looks over a few times and one of my girlfriends notices. 'What's going on with you and him?' she asks. 'Nothing,' I say. He is wearing a grey T-shirt and I notice his tanned muscular arms with scratches on them, fresh from the sunshine and work in the Australian Outback. His hands are cracked and rough. His friends cluster around him to hear his stories about kangaroos and picking pineapples, but he doesn't want to talk much about being away. He seems happy being back amongst them, laughing and joking. None of the lads have girlfriends so they all go out as a big group every weekend. I've seen them from afar before, all dressed up in their dads' 1970s suits just for a laugh. Then the attention turns to one of his mates, who runs into the pub to tell the group the 'coppers' are taking a lad they all know into the station for fighting.

Everyone spills out of the pub to see what's going on. James is holding a bottle of beer and I'm standing on my own for a moment, my glass empty. He asks if I want another and I nod: 'OK.' As he goes back to the bar, the group drifts on to the next pub. I say to my friends, 'I'll catch up with you soon, after this drink', and follow him inside. Away from everyone else, we chat about his trip and he asks how sixth form is going. I thank him for the postcard.

We eventually follow the crowd to the nightclub. Everyone else is pretty wasted, the girls mostly dancing and the lads ogling them. I ask if he wants some fresh air, and we linger outside for a while and kiss at the side of the club.

Helen comes out to find me. She's got us a taxi because she promised my mum that we wouldn't be late – I have work tomorrow. She laughs at me all the way home and says, 'You said you didn't like farmers.'

'Yes, but he's different,' I say, smiling.


### Basket of Chips

James picks me up the next week and we drive to a country pub that his friends don't go to. He teaches me how to play pool and we share a basket of chips that he orders just before they close the kitchen. When he tells me he read *Wuthering Heights* on his travels because I had told him I was studying it for A-level, I know I want to be with him more than anything else. He drives me home along the windy country lanes late at night.

We go for walks together at weekends, hiking up the Horse Pasture or along the Long Meadow on his farm. I ride alongside him in the tractor when he has to work. He shows me birds' nests in the hedgerows and takes me to see salmon spawning in the becks at his grandad's farm, with torches in the dark. He reads Russian poetry to me and pretends to watch all the French art films with subtitles that I want to. We lie together on the velour sofa that has no springs left in it, in front of the TV screen, late at night at his parents' house. We spend most of the time ignoring the film. His sister doesn't speak to me much after that.

We plan to go away for a weekend. He suggests Scotland and borrows his parents' car. We walk around Edinburgh eating ice creams on a bitterly cold day and spend a night in a hotel before driving north as far as Inverness and staying in a tiny B&B at Drumnadrochit. In the morning we sit down at the neatly laid table with cups and saucers set out. The owner pours tea into my cup, scowling as she walks away. James gazes out the window at Loch Ness. The water is still, the mountains reflected in it. He is completely unaware of her disapproving looks towards me. She places two cooked breakfasts down in front of us and doesn't smile, clearly thinking to herself that I am too young to be away like this.

For James's twenty-first birthday I write out some of his favourite poems and paint little pictures of brambles and swallows in a ring-bound notebook with a green velvet cover. I know I want to be with him for the rest of my life. I try to impress him with my cooking; I don't want him to think all we eat at home is stew and mash, so one time when my parents are away I invite him over for a meal.

This first meal I cook for us is a Spanish chicken stew, with red onions, peppers, paprika and tomatoes. To my eyes it looks delicious. I don't add olives because buying a jar to use only a few seems extravagant. The table is set with a vase of sweet peas cut from the garden, there is a basket of fresh bread and a bowl of steaming rice, and then I lift the lid of the pot full of delicious stew. I ask if he wants to help himself and I sense relief as he quickly spoons a tiny portion onto his plate. 'Aren't you hungry?' I say, and he mumbles something and then helps himself to more bread, getting busy buttering it. After we have eaten, he rushes to gather up the plates before I offer him any more and helps me wash up.

It's not until a few weeks later that I work out the closest thing to flavour he has ever tasted is probably a packet of prawn cocktail crisps. He hasn't told me that he mostly eats cheese sandwiches and chips. He's a very fussy eater. We go out for meals together and gradually I get him to try some of mine, and bit by bit he becomes more adventurous – I suspect out of fear of not wanting to seem ignorant.

## SPANISH CHICKEN STEW

Prep 15 minutes
Cook 1 hour 10 minutes

Serves 4

### Ingredients
olive oil for frying
500g/1lb chicken thighs/breasts
50g/2oz chorizo, cut into chunks
5 garlic cloves, minced
2 onions, sliced
1 red, green or yellow bell pepper, sliced (or a mix)
1 head of fennel, trimmed, or 1 courgette, chopped (or both)
400g/14oz tomato passata or a tin of chopped tomatoes
1 tbsp tomato puree
a few sprigs of fresh thyme or 1 tsp dried thyme

400g tin of cannellini or butter beans, drained

1 tsp sweet paprika

280ml/½ pint chicken stock

8 pitted green or black olives, whole or halved (optional)

salt and pepper to taste

balsamic or sherry vinegar to serve

fresh herbs to serve

Note: If you don't have fresh bell pepper, use some roasted and preserved pepper from a jar. Also, if you have some white wine to hand, halve the stock and add a large glass at step 4.

## Method

1. Heat the oven to 180°C/fan 160°C/gas 4, unless using a slow cooker or the hob. Heat a little olive oil in a large, heavy frying pan or casserole-type pan with a lid over a medium-high heat, and fry the chicken pieces until golden. Don't overload the pan; do it in two batches if necessary. Add the chorizo at the end to release the flavour.

2. Lift the browned chicken and chorizo onto a plate while you cook the vegetables.

3. Add in the minced garlic, onion, sliced pepper and fennel/courgette and cook for 10 minutes, until the vegetables are softened.

4. Add the tomato passata (or tin of tomatoes), tomato puree, thyme, beans, paprika, chicken stock and olives if using. Season well with salt and pepper.

5. Return the chicken pieces and chorizo to the pan and put the lid on. Cook in the oven or over a low heat on the hob for about 40–50 minutes. If using a slow cooker, cook for 2 hours on low.

6. Finish the whole dish with a little drizzle of balsamic or sherry vinegar and a scattering of whatever fresh herb you have to hand; parsley, coriander or more thyme work well.

My mum loves James too (or sees a way to get rid of me). She chats to him about the headlines or a random news story she's read in the *Daily Mail*, and he always knows what she's on about. She tries to impress him by making puddings every time he comes over for a family meal – apple crumble, rice pudding, lemon surprise or sticky toffee pudding. My dad is happy about this because he doesn't otherwise get puddings. Dad talks to him about sheep and cattle and what prices they are making at the auction. But I am the only one he talks to about books. Only I know he writes poems and stories and is always reading.

I leave school and start my foundation art course at a local college. I am nineteen and desperate to travel. I want to be away from home as much as possible. I've been on art A-level trips to Florence and Venice and I'm reading *Under the Tuscan Sun*. I imagine myself living in a villa surrounded by olive trees instead of here in the attic of a cold farmhouse. When I spot an advert in a newspaper for a cheap Italian holiday apartment with a phone number, I look the town up in the atlas we keep in the living-room cupboard and discover it is in the heart of Tuscany. I phone the number and it all sounds too good to be true, so I suggest some dates and promise to send a cheque for the stay. I book us flights though Thomas Cook in town.

When we arrive in Florence, we figure out how to get train tickets with no word of Italian between us. We take the train to Montepulciano and then hire a tiny car that is more trouble than it's worth – it overheats on the steep hills. We find the address with the vague directions I've been sent in a letter. The rusty iron gates open up to a long cypress-lined drive. There under a stone, as promised, is a big old key and a handwritten note. Our instructions are not to stay in the little apartment that we booked in the main house – it is having work done – but to use the cottage adjacent instead, where there is fresh bed linen in the wardrobe. I can't believe our luck – the whole cottage to ourselves. I wander around the cool rooms and check out the kitchen, which opens out onto a vine-covered terrace with a marble dining table. James strips off and jumps straight into the pool to cool down after our journey.

It isn't until the middle of our two-week stay that things start to go wrong. My family have always taken an annual holiday in May. We've been 'Brits abroad', playing endless card games in rainy Majorca or Tenerife. Dad once rented a villa in the Algarve from a family friend and we went to the Slide & Splash water park for my sister. We've often taken shorter trips to Fort William and the Isle of Man too. I know what to do on holiday. But James's family couldn't afford to take holidays. His work trip to Australia was hot, dusty and dirty and he didn't enjoy it very much. This holiday in Tuscany is only his third time away from home. I'm expecting him to know how to relax and take in the sights, slow down and enjoy nice food, because the whole place is my idea of paradise, but he has had enough of this imposed leisure time after a week. Especially as he is swiftly getting through the pile of books he's brought (packed instead of clothes). We don't go far because of the crappy car. He spends hours walking or reading. I spend the long days keeping out of the burning sun with my red hair and pale skin. I swim, draw, paint and walk to the market every day to buy food for us to eat at the marble table under the vines. Tomatoes, salad, charcuterie, pastries, bread and fruit. Simple fresh food that tastes nothing like it does at home. In Italy that summer I really get to know him. He is uneasy on holiday because his life revolves around the relentless work of the farm. He finds doing nothing hard. My life is finally starting, I love art college, but he is deeply unsettled. He can't see a future for himself on his family farm, but it's the only thing he knows how to do. He feels frustrated. Some days I'm not sure I feature much in his mind, and he tries to push me away because he thinks we want different things. But he comes back to me when I ask him to read parts of the books he is reading. Bits from Albert Camus' *The First Man* or poems by Anna Akhmatova, read aloud to me in the evenings. I lie looking at the tiled ceiling of the living room, bathed in the evening sunshine, and we drink

Chianti. He loves telling me the stories from *And Quiet Flows the Don* by Mikhail Sholokhov.

We take the train into Florence. I sit watching him reading on the journey, lost in another story. I stare out at the countryside, every view postcard-worthy. We are going into the city to see if we can watch Princess Diana's funeral on a screen somewhere. I phoned Mum from the village square the other day after a British couple heard us chatting in English and interrupted us to ask if we'd heard the news: Princess Diana had been killed in a car crash in Paris. Mum filled me in a bit more. 'Britain is in shock, everything is at a standstill. People can't believe it. Those poor boys . . .' she said. All the bars with TVs are crowded with people trying to get a glimpse of the funeral. We walk into a large book-shop to see if there's an English-language section to keep James going. Inside, there is a huge screen near the tills and I stand and watch the procession and the royal family following behind Diana's coffin. After a while we are back outside, in amongst the swarms of people. I'm struggling in the heat and want to find a cool gallery or church to retreat into, but the queues for the Uffizi are ridiculous and we can't get near the Piazza del Duomo. I see that James is struggling too. He hates the crowds, the city streets and being far away from the green fields of home. We take the train back, not speaking much. I had wanted to go, to feel the buzz of the city and have a break from our very quiet first week together. I thought it would do us good, sensing that he was unhappy, but I know he needs the pool and some space.

That evening he tries to break up with me. I listen quietly as he tells me I will be better off without him. He wants to change his life but is struggling to work out how. He loves the farm and his family, but it has gone a bit toxic. He doesn't have any qualifica-tions or other skills to get another job and feels totally lost. Most of all, he doesn't want to hold me back. We talk late into the night. I know I see something in him that no one else does. I am not going to give up on him and I am not going to let him push me away. I love him.


## Water

My neck is stiff. I can barely turn it. I feel so tired. My eyes are open as far as I dare. I blink as we enter a little room with a bed. I gently turn my body away from the big window overlooking the hospital car park. Mum closes the blinds and pulls the curtains shut without me asking. I kick my boots off and climb up onto the bed, pulling myself with the metal handrail. I lie on top of the neatly folded covers, curl into a ball and hide my face in my coat. The nurse comes into the room, her rubber shoes squeaking on the floor, and goes to turn on the lights. Mum quickly stops her and tells her I am struggling with the light. The nurse takes my temperature again and gives me some pills to swallow. 'You'll be OK, my dear,' she says. 'The doctor will be along to see you soon.' I want to cry.

Mum sits upright by my side, alert for anything. We don't speak.

My head throbs and I feel faint after throwing up a few times before we got here. Mum called our local surgery and got me an emergency appointment. She seemed frustrated with me and said it was probably a migraine and that I should stay in bed, but took me in anyway. I said I would be fine, so I went in without her. Mum went off in the car to get some shopping. After I'd been checked over by the doctor, I waited for her on a bench by the door, turning my head away to ignore the passers-by, because there is always someone familiar at the surgery in our small town.

My body was shivering and shaking and I felt awful. When Mum finally drove up to the door I held back my tears to speak to her. She lowered the window to say something but I got in first. 'Where have you been?' I said sharply. 'I have a temperature. The doctor says I have to go straight to Carlisle hospital.' She tried to speak again, but I wasn't in the mood for how busy the supermarket was. I shut her down: 'He said it might be meningitis. Do you want to take me or should they send me in an ambulance?'

As soon as I said the word 'meningitis' she immediately switched into helping me. She jumped out of the driver's seat and moved the groceries off the front seat into the boot, and I got into the car. She drove fast up the motorway, right to the front of the big city hospital, and demanded that we get some help at the reception desk. Her voice was firm, but I could sense fear in it. The fear seemed like it was for me, and that felt strange. She helped me through the corridors, following a porter who swiftly came to show us where to go. Mum had taken charge, and my usual need to speak for myself vanished.

Now, I lie still as the doctor tells her that they want to give me some more pills and a lumbar puncture. Not long after the doctor has gone the door opens again, and James walks in, whispering 'Hello' to Mum. He walks straight to me and kisses my head for a long moment. I mutter, 'You stink', and I see he is in his farm clothes, his jumper covered in wisps of hay and his boots caked in sheep shit. He takes his coat off, casts more hay onto the floor and laughs. 'I'm glad you haven't lost your sense of smell,' he says.

We wait. My shivering has stopped but my head is still pounding, and now I can't stand any noise as well as the light. James and Mum chat quietly and look anxious. The nurse comes back in to say, 'We are going to wait until tomorrow to do the lumbar puncture. Helen will be staying with us overnight and you can go home. We'll ring you if anything changes.'

James shows me a bag he picked up from my home on his way, with my toothbrush, pyjamas and a book from my bedside table, an art history guide. I smile and say, 'I don't think I'll be doing much reading.' I can see the worry in his eyes. Loving someone like this is new for him, and it scares him, but he is here.

They gather up their coats quietly. Mum passes me a glass of water and asks if I need anything else. I feel uncomfortable with her fussing and say the nurse will get me anything I need. I lie still and go over it all in my head. I've seen a different Mum today to the Mum I thought I knew, the Mum I have argued and fought with for ages. In those scary moments I saw another part of her that has been hidden from me for a long time. I overheard her talking with the nurse on her way out. She told her that, when she was six, her baby sister died of meningitis; she was fourteen months old.

The next morning, after a deep sleep, I know I am going to be OK. The room is dark and quiet and tucked away from the busy corridors of the hospital. I can move my neck freely and my head feels clear. I gently climb out of bed and stretch my arms up over my head. I open the curtains, look directly into the light and let the sunshine warm my face. I take another moment to realise I am alright. Then I wander out of my room and along to the nurses' station. I talk to a different nurse and she says once I have been seen by the doctor again I will probably be discharged, and that she will ring my mum. The doctor is on her rounds soon and when she gets to me she checks me over and tells me I have probably had viral meningitis.

My world has changed in the last twenty-four hours. I know that James is here for me, that he loves me. And I can feel the last few years of angst between me and Mum melting away. She would never say it out loud, but I know she loves me. I am no longer a girl and she is adjusting to that. For the first time, too, I can see my mum as a person in her own right. Yesterday I overheard her speak about something she has never told me, and I want to work out how to ask her about her past.


### Porridge

Breaking a mirror is supposed to bring you seven years of bad luck, but I have decided I'm not superstitious, and besides, I'm at art college. I need some shards of mirror to make a piece of work about memory.

I put the cheap glass face down in a tea towel and hit it cautiously with a hammer. I hear a crack on the second, firmer bash and know it has broken. Unfolding the cloth carefully with my fingers protected in a pair of old leather gloves, I lift the largest pieces onto a sheet of black mountboard. I have a pile of old black-and-white photographs on the table. Pictures of me and my mum as little girls, thirty years apart. One of my mum pushing a doll's pram on a lane. And one of me on a swing in the garden, laughing. I place the angular pieces of mirror over the pictures and glue them all into place to make a montage; it looks as if the mirror has shattered over the images. I hold it up later when the glue has dried, and I see myself in the fragments of mirror amongst the photographs.

When I started the project I asked Mum questions about the photographs I'd found in the living-room cupboard. Bit by bit she told me stories as we looked at them, and for the first time I pieced together who she was, and is.

I picked up a small square photo of her with short bobbed dark hair, holding a black cat quite tightly in front of a little white-washed building on a lane. She said the photograph had been taken at her great-aunts' farm. She was seven or eight years old.

Mum had three great-aunts – her dad's mother's sisters. They'd brought her dad up after his mum died when he was a baby. These three eccentric spinsters lived together on their farm. They kept a lot of hens, to sell the eggs, and rented their sixty acres of land to a neighbouring farmer. They didn't have electricity and chose not to get it many years later when it was offered to them. Mum stayed there most weekends and holidays. She had to take a Tilley lamp filled with paraffin up to bed. Her days were spent roaming around all over the farm with an old collie dog. She told me she loved poking about in the streams – or 'burns', as they called them. One aunt regularly made her drop scones (little pancakes with butter) on the range.

Her dad was always busy working at the bakery, because he wanted to buy the house that came with the job. He'd been a cook in the merchant navy during the war and would tell her tales of his life at sea. Mum smiled when she told me that he always brought her delicious warm bread and cakes, even doughnuts on a Saturday. When she married my dad, her father baked the wedding cake.

When she was little, her dad would scoop her up and put her in front of him on his motorbike, and off they'd go together. But she had two younger brothers, and they started wanting rides too, and there wasn't room for them all. If they left her behind, there wasn't much for her to do at home, so she would walk the three miles to her great-aunts' farm. One day, walking home in the late afternoon, she heard a rustling in the woods along the lane and it scared her. She heard heavy breathing and pounding sounds on the ground behind her as she ran. She was tumbling in her panic, and when she turned she realised that the cattle had got into the woods and were charging around trying to escape. Big blundering shapes galloping around in the shadows. She ran all the way home and never walked back alone in the dark again.

Some Sundays, her dad would stay and dig the garden for his aunts or help gather fruit from the bushes and trees in the orchard. Aunt Nell told my mum that she had to help in the house when she was there. She used to muck out the hen huts, collect the eggs and feed the hens with a bucket of 'mash', which was all the waste peelings from the house chopped up and cooked in the bottom of the stove. But Mum noticed that her brothers never had to do any chores; they were allowed to play. One time she called the aunts 'mean old buggers' because they made her do all the washing-up, and one of her brothers told on her. She was punished with more chores, but she said it was still better than being at home. I asked her why her mum let her go there so much, and she said, 'Oh, she didn't notice what I did', and turned to get on with emptying the sink.

I knew that Mum loved and was loved by her dad – she always spoke of him fondly – but when it turned to any talk of her mum, my gran, she always shut the conversation down.

When Mum was seventeen, her best friend Christine wanted to leave for the city. Mum wasn't really for going, but it came to the crunch one day when Christine said, 'Well, I'm off, are you coming? We'll have loads of fun.' Mum said to me, 'I went home after that conversation, I looked around and figured there wasn't much to stop at home for. Being there wasn't living – it was starting to get me down. You start to think the things they say about you are maybe true, so I thought "Stuff this", and then the fight came out in me. I packed my bags, went to meet her and got on the bus.' They rented a room in a big house in Carlisle; years later she would point it out to me when we went shopping. They did a couple of days a week at college and then worked in a hair salon – doing perms, blue rinses and shampoo-and-sets for the old ladies. The photographs of their nights out show them all dressed up in knee-length platform boots with frilly blouses and short cheesecloth skirts, hair all puffed up with hairspray, and thick black eyeliner. They went to The Cosmo on a Saturday night and drank Babycham, and danced to The Who, The Troggs and Bay City Rollers. One night they went along to a 'Young Farmers' dance and Mum met my dad. They started regularly hanging out together at The Cosmo; soon all of their friends were paired up, and one night they got together.

One Saturday I persuaded James to come with me to visit Mum's hometown, Lochmaben, in the Scottish Borders. I wanted to see where she had lived. My memories of visiting this house were blurred because I had been so small. Mum's dad died when I was three years old, and with that all her ties to her old home were severed. My gran still lived there, but we had no contact with her. We found the house at the end of a road in the bleak little town. The paint was peeling from the sandstone windows. The bakery her dad had worked in was long gone. In my memories the house was much bigger. Mum, my little brother and I had walked through that front door, years ago, and sat on a firm green sofa. A white porcelain dog stared down at me from the mantelpiece in the front room. This thin, frail old lady had a crooked back and looked like she might snap if she stood upright. She tried to be kind to me, but there was something wrong in the room between Mum and her. She had wispy white hair and wore light-grey nylon trousers and a pale-pink cardigan over a blouse. She moved slowly, hunched over with a stick, into the little kitchen and slid open the rippled-glass doors of a cupboard on the wall. It was full of packets of biscuits. There wasn't any other food around. She sliced the plastic of one of the packets through with a sharp knife and passed me a plate to put some custard creams on. I stared at her long bony fingers. She made us kids a glass of orange squash each that was too strong. Mum was pregnant with my little sister and we didn't stay long. On the way home, Mum took us to Dumfries to a cafe that she knew served pancakes, and I had one called Cherries Jubilee off the menu, with hot black cherries in a gooey syrup and a fast-melting yellow block of vanilla ice cream in the middle. I loved the hot-and-cold taste of it all and cleaned the plate quickly with my finger before Mum told me off.

My gran sent me a letter soon after that, but when I opened it Mum snatched it off me, saying there was nothing she had to say that was of any interest to me. I remembered the letter when we were talking about the photographs; Mum just said, 'She tried to poison you against me', and that she had written horrible things about her. My gran claimed that it wasn't her fault my mum didn't want to go and visit, and she wrote that she wanted to see us more. I never saw that letter again and we never went back.

It was raining, and I had no intention of visiting the old woman. I felt a fierce loyalty to Mum. James and I parked by the gates to the cemetery on the edge of town and we walked around until I found the names I was looking for. There they were, scribed on a plain grey headstone in simple writing. Rain was trickling down the letters. My grandfather, James Thomson, and my mum's little sister, Helen Thomson. I read aloud the dates of her birth and, so soon afterwards, her death. I gasped. I felt crushed with sadness. I had been named after this baby. James held my hand as the wind blew against my face. I hadn't noticed I'd got so wet from the fine misty rain, and we hurried back to the car.

Later that week I was drawing up a family tree and wanted to know the dates and names of all of Mum's side of the family. It was easy to get information from Grandma on my dad's side; she would talk for hours about everyone and the farms they lived on. But Mum was the only person I could ask for the names on her side. She helped me fill in all the names of her aunts and cousins and told me who they married, and when I came to note her and her brothers' dates of births, she told me more about the death of her baby sister. She didn't know I'd just been to visit her grave.

'I didn't go to her funeral,' she said, quite matter-of-fact. 'Children back then didn't go to funerals.' Her mum's neighbour, Mrs Harvey, had looked after her that day and made her some dinner. Later, when she went back home, her house was full of people she didn't know, and no one really spoke to her. After that day, she said she just played with her dolls and kept herself to herself. By the age of about twelve she was doing all her own washing and often fed herself too. Her mother told her she frowned too much, that she was a 'miserable child' and 'not at all like your little sister'. 'I didn't answer back much,' Mum said. 'I just chose to be quiet from that day onwards.' But she added, 'She made some things I liked – mince and dumplings, and braising steak with onions and mashed potatoes . . .', pausing as she remembered, '. . . but she never showed me how to cook anything.' One day her dad brought a chicken back from the farm and chopped its head off in front of her, so she said she went off meat for a while after that.

When Mum was a girl, one of the big news stories in the papers was about the Moors Murders. Ian Brady and Myra Hindley had abducted and killed children on the moors around Manchester. She had read bits about the case and couldn't sleep. But when she told her Mum this, my gran's reply was simply: 'Well, when it's your time, it's your time.'

Mum's voice wasn't emotional recalling these parts of her child-hood, there was no anger or frustration at the way her mum had been, but as she said these things I noticed a pattern. She always made sure to add a happy detail, as if I would think it all too bleak otherwise. I sensed that if she dwelled on it all too much it might break her.

Mum told me about one time they went shopping together and she saw a little porcelain angel on a shelf that they both liked, so her mum bought it for her. She said it would remind her of her little sister. Another day, she bought her a piggy bank in the post office with her name, Elizabeth, painted on it. Listening to her, I felt an overwhelming sadness for that little girl who was my mum, and another kind of sadness for her mother. I asked her if my gran had ever been diagnosed with depression or got help in any way. She said there was no understanding of things like that, like there was now. 'You can't fight depression, it's like being an alcoholic, but she was her own worst enemy – she wouldn't have help.' And she added quietly, 'It was all very sad.'

I wondered if anyone had ever properly hugged Mum when she was a little girl. My 'auntie' Christine helped me fill in a few more gaps on the phone one day. She said my gran barely acknowledged Mum's existence after her baby died. Mum had been a spirited six-year-old until then, but suddenly she had to learn how to manage on her own. Christine called on her one day to see if she was alright because she hadn't been in school – she thought they must have been about eleven years old – and my gran opened the door and said sharply, 'Of course she's been in school, she's fine.' But it turned out that Mum hadn't been well, and had stayed in bed all day, and my gran hadn't even noticed. Mum figured out ways to get through those neglected days. She helped her dad fix his motorbike and spent more and more time at the farm. Christine's house down the road was a sanctuary. Her mother, Mrs Kennedy, was kind to Mum, and later became the Scottish gran to me that I didn't have. I remember staying at Christine's house and playing with her two girls when I was little. Mrs Kennedy was regularly there, and she would let us do her make-up with our face paints and style her hair; she made us all laugh so much. Remembering her warmth to me as a child makes me glad that she was around for Mum.

Mum's best childhood stories were always from her aunts' farm, chasing animals around, digging up potatoes, cooking porridge in a black pan over a fire and going for adventures on her own. She told me she stood on the hill one day, when she was ten years old, on their farm, and she looked out and said to herself: 'Someday I'm going to marry a farmer, because that's the only way of life that has any meaning.' She told Christine, who said, 'Ugh, I'm not marrying a smelly old farmer.' And Mum thought, 'No, I bet you don't either.'

## PORRIDGE

### CLASSIC PORRIDGE

Serve your porridge with whatever toppings you prefer. I like maple syrup and blueberries. Dried fruits like raisins or sultanas, stewed plums, or apple and cinnamon go well with porridge too.

**Cook 10 minutes**

Serves 1

**Ingredients**
50g/2oz rolled oats
100ml/4fl oz water or milk (or half water and half milk)
a pinch of salt if you like

## Method

1.  Put the oats in a pan with the water/milk (and salt if using).
2.  Cook over a medium-low heat, stirring for 10 minutes until thickened.
3.  Spoon into a bowl to serve. If you like your porridge looser, add more liquid and stir through on the heat.

## OVERNIGHT OATS

This summer alternative is a basic recipe that can be adapted to suit your taste.

**Prep 10 minutes**

Makes 1 generous serving

## Ingredients

75g/3oz rolled oats
2 tsp chia seeds
½ tsp vanilla extract
1 tsp maple syrup
1 tbsp Greek yogurt
150ml/5fl oz milk
½ grated apple

## Method

1.  Stir all the ingredients together into a large jar with a lid, or Tupperware, cover and leave overnight before adding any additional fruits, spices or seeds you like on your oats. Homegrown blueberries and raspberries are my favourites.

These conversations Mum and I had at the start of my art-college days were life-changing for me. They transformed the way I saw her, and also the way I saw myself. It was clear that in her late teens, Mum had made a choice about how to live. She had experienced life with a mother who barely had a kind word to say to her. A mother who neglected her. A mother who had almost dragged her down. Yet, looking forward, she chose to live a more hopeful life. She took the best parts of her childhood and thought about those instead of the worst; she went out into the world and chose to be happy.

Mum is tough and there is a fire inside her. She has not let her past dominate her head or her life. I think she decided a long time ago that she is her own best bet. I saw the fire in her regularly when she was dealing with my grumpy grandfather, or when she and I were arguing about some minor thing in the house. But it's mostly dormant now. She occasionally gets angry with the world if she ever feels like she is being controlled or bossed around. She works hard but always does things on her own terms. Her way of loving me is to do all the little unseen things to help me. She comes to our house and rolls up her sleeves, tackling the pile of washing-up in the sink or the weeds in the garden.

She said to me one day, 'When your own family doesn't love you, you have to go out and make a new one.'

# MORNING

I have only just got the breakfast dishes cleaned up and hoovered the kitchen floor when James is at the door. 'Do you want to come and see if Heidi has calved?' he asks. I leave the pile of unopened post, my next job, on the kitchen island, and climb on the back of the quad bike with him.

Down the field, Heidi is lying on her side, a few feet away from the herd. The other cows are settled and chewing their cud, flicking flies away with their tails. Two calves, born three weeks ago, are tucked amongst them, the cows around them like a herd of protective elephants. Heidi is straining with regular contractions and I can see two hooves poking out.

'Can you see?' James says, and just as he stops the engine, 'Yes, yes, the legs are out!' I say a little too loudly and excitedly. 'Sshhh . . .' he whispers. I know before he says it aloud that we will be here until the calf is born. 'We'll wait, it might be another twenty minutes or so,' he says. 'That's OK, I was only tidying up.'

Early this morning he spotted that Heidi was showing signs of calving. I think he's been back to check on her several times since then. The field has a stream running down one side of it, and the grass is long and forms gentle waves in the breeze. White pignut flowers speckle the meadow like confetti. James points out a roe deer in the distance.

It isn't long until the head of the calf emerges, with the front legs in a diving position coming out from its mother. Heidi struggles to her feet, with the calf half out, and uses gravity to help her expel the lithe wet body. It tumbles onto the grass and kicks its legs. I want us to rush over and check it is OK, but James holds back. 'Wait!' he says. 'It has lifted its head up – it's fine.'

Heidi turns around to see what has just appeared out of her. She starts to lick the birth sac away from around the new calf's face and body, and moos gently around it. We drive a little closer. The whole herd has noticed the new arrival; they are up and making their way over to see. Dinky moos and calls to her own calf and she gallops over to her. Cara and Eyebright are the most curious and maternal – they are due to calve soon. Heidi stands over her newborn, a proud mother, and all the aunties fuss around it as it tries to get to its feet.

'Wow, that's fast,' I say, and James looks under its back end as closely as he dares.

'It looks like a bull calf from here.' I can feel the relief in his body, relief that he hasn't needed to intervene. He is smiling.

I put my arms around his waist and squeeze. 'Brilliant,' I say, and he leans back to give me a kiss.

He starts the engine and we head back to the house. There are lots of tough days on the farm when everything goes against you, but on a morning like this there is nowhere else on earth I'd rather be.

James kicks his boots off by the door and flops in a heap in the chair, tired from worrying about all the things that could have gone wrong with the calving. 'Tea?' I say, and I put the kettle on. I can hear Tom and Mum upstairs, chattering about getting a towel and some dry pants. There is an abandoned hosepipe on the patio and a puddle of Tom's wet clothes on the floor in the doorway.

'What do you want to eat?' I ask James.

'What is there?' he says.

'Well, I've got some tomato soup, or a bit of cold sausage and some cheese and bread.'

'Just whatever,' he replies.

I warm the soup in a pan. Tom will have some too. It has a few extra vegetables and a handful of red lentils blended into it, and

I stir a blob of cream in as I serve it. My four-year-old tornado of a boy charges into the kitchen laughing and tells me all about 'accidentally' showering himself with the water pipe when he got back from nursery, and I act surprised at his mishap. I set out some plates on the table and unwrap the cheese. We butter thick slices of bread from a sourdough loaf I bought yesterday; it's a bit dry but OK with soup. Mum and I talk about the dreadful news from Ukraine. James clears the plates away and I load the dishwasher.

Afterwards I take a basket and swap the washing over on the line strung between two trees. Mum is busy weeding the garden and Tom is driving his toy tractor around the patio, gathering loads of her weeds into his trailer. A wood pigeon coos up in the oak tree.

James goes to check on a poorly sheep in the barn, and to do another shepherding round. I call the accountant, apologising for missing him earlier. Then I go and find the pile of unopened post. I don't have a very good system for incoming paperwork – I generally ignore it until I have to deal with it. James's mum did the farm paperwork for many years, but when we moved here she arrived one day carrying several plastic boxes of files and folders, all stuffed with papers. She pretty much said to me, 'Here, it's all yours now!' with a sigh of relief. I look forward to the day when I am no longer responsible for filling in medicine books, doing movement licences, registering cattle passports and keeping all the relevant farm assurance inspection documents. Paying the bills is about as much as I can squeeze into my days sometimes. The annual insurance and accountant meetings fill me with dread; I fear they will catch me out with something and discover that I really have no idea what I'm doing.

I make a cup of coffee, open my laptop on the kitchen island and look at my emails. 'You have six new messages waiting from the Rural Payments Agency, please log in to read.' But just as I am about to find my login codes, Tom runs up to me with a plastic tub of water he has carried in, filled with soil and stones. 'This is going to be my lava pool!' he says, and I shout, 'No! Not on the carpet!' He howls at my sharp reaction. I try to take the tub off him before he spills the mud everywhere. 'Here, take it back out to Granny, she'll help you make a lava pool in the garden. You can take some dinosaurs outside.' My voice softer, his face quickly lights up again, as if this is his idea.

James has just pulled into the driveway. 'Can you come and give me a hand tagging lambs?' he shouts. 'It won't take long, there's not many in this batch.' Any excuse to avoid the paperwork, I think, and I push it to one side.

I check with Mum that she is OK to stay with Tom. 'We're fine. I can stay until you go for Isaac at three.'

Most of my days go like this. A good friend told me about a French story she'd read in which a woman grew several arms to deal with all the tasks that came her way. I laughed at the thought, but I felt the sad truth behind that tale, a desire to be able to cope. My roles as a mother and a farmer's wife demand a lot of me. Most of the time I am OK with it – I choose this life. But sometimes it is overwhelming. I regularly feel I have too many things that I am responsible for. My mind gets crowded as the jobs compete to become my priority. I just end up doing the most urgent thing in front of me. I feel like I don't do anything very well, as I can't really give anything my undivided attention.

Sometimes the children come first. 'I want to paint, Mum!' 'Can I have a snack?' 'Come and see this, Mum.' 'Read this to me!' The trainers are too small, someone has a swimming lesson, someone needs a lift to work, the school needs a form filling in. Have they done their homework? Who left the Play-Doh out? 'Can you renew my bus pass?' The games need to be played, the colouring needs to be coloured in. Someone has left muddy boots in the doorway, someone has lost a shoe, someone is crying, someone is laughing, someone is sulking, someone needs a hug. Someone is hungry. Someone is always bloody hungry.

Sometimes the farm comes first. There is a sheep on the road. A lamb stuck in a fence. The hens need water, the ponies need shoeing, the cows need their electric fence moving. 'Can you come and move some sheep?' The petrol cans need filling from town, the shed needs sweeping, the sink needs unblocking, the dogs need exercising, the kennels need mucking out. 'Have you phoned the knackerman to come?' The workshop needs cleaning up, the gate needs shutting, the new calf needs tagging, the diesel tank needs checking. The eggs need collecting. The bloody eggs always need collecting.

Sometimes the paperwork has to come first. The contract needs signing, the date of the event needs confirming, the invoices need sending, the car needs servicing, the insurance needs checking, the medicine records need updating, the calf passport needs sending off, the trees need ordering, the health and safety forms need reviewing, the report needs reading, the accountant needs papers signing. The bills need paying. The sodding bills always need paying.

Sometimes the housework comes first. The floor needs hoovering, the washing machine needs emptying, the wet things need hanging out, the dry things need gathering up, the mountain of folded clothes needs putting away, the fridge needs cleaning, the shopping needs unpacking, the carrots are wilting. The table needs wiping, the plates need stacking, the leeks need rinsing, the potatoes need peeling. 'Is there any bacon?' The cobwebs on the landing need brushing away, the shower screen needs fixing, the washing on the line outside needs fetching in before it rains, the hosepipe needs gathering up, the shoes and boots need tidying by the door. 'What's for supper?' The pyjamas on the stairs need taking up, the towels on the floor need hanging up, there is Lego down the side of the sofa, there is a stain on the carpet, the ash in the stove needs emptying, the books on the mantelpiece need putting away, the toilet roll has run out. And in my manic head, the cushions need straightening. The fucking cushions always need straightening.

Some days I feel like I am slowly drowning in this life.

When I'm really tired, I go about my jobs grumbling under my breath, building resentment and then muttering my frustrations aloud. 'Who has left this empty cup here?' or 'Why have you not put that away?' I hate my nagging voice. That voice tells me I have neglected myself for too long and need to find some time to do something on my own. When it gets like that, I basically feel sorry for myself. I believe that everyone around me has it better than I do. I convince myself that their day doesn't involve any of this relentless picking-up, that I'm juggling all their needs and wants so that they can go about their day. No one needs anything from them. But I know this is a myth. Their days are every bit as relentless and sometimes gruelling, just in different ways. But on the bad days, I want to scream at everyone, lie down in a dark room or run away.

The mental load of the family and the home seems invisible to everyone else. I often try too hard to make everyone around me happy, but I have realised over time that in a family of six it is going to be very rare that we are all content at once. I can only do my best and I must try to take care of myself too. I often just need to get outside and breathe.

# 2

*Pizza*

We are driving through the city, car boot wide open, back seats folded flat, and a long fridge-freezer hanging about a metre out of the car. Baler twine is strung around the bottom of the unit to stop it falling out. 'We're nearly there,' I say, relieved that the police haven't stopped us. James yanks into reverse and backs in as if he is driving the farm pickup with a load of hay on it. I jump out, open the front door and run downstairs into the basement flat for the scissors because, despite having a bundle of twine in our car boot, James no longer carries a farm penknife in his pocket. I prop the door open and move the boxes out of the way, making a clear path. We nudge and swear and wrestle it up the back steps, round the corner. 'Careful, watch the lights . . . Yes, just one more step . . .'

The door of the ground-floor flat opens just an inch or two and an elderly lady's face peers out at us. She is white with face cream, and she startles me. We stand the fridge upright and take a moment, swapping places so that James can go first down the stairs, taking most of the weight. I smile at her and prepare to say 'Hello!' as I manoeuvre around the banister to lift this beast of a fridge again, but she shuts her door quickly. As we get it into the flat, my arms are shaking. We set it down and then burst out laughing, shuffling the tall fridge on its wheels along the thin carpet into the narrow space at the back of the living room. The floor has one of those metal runners between the carpet and a little patch of lino, and as we lift it gently over it glides into place between the sink,

Baby Belling cooker and a tall cupboard. There is a faded-navy velvet curtain to pull across the alcove that hides this space from the rest of the living room. This isn't what anyone would call a proper kitchen, but I don't care – it is mine and I love it.

We arrived in Oxford yesterday with all our things, box after heavy box of books, bags of clothes and a whole heap of bedding and kitchen utensils. We were up and down those stairs a hundred times unloading the hire van carrying the pine double bed frame, its mattress, the small wooden chest of drawers that I'd painted cream and the green tatty second-hand sofa that I plan to cover with a stylish throw from Habitat. I hugged James, beaming inside and out that our life together was truly beginning. We were on our own, away from the interference of our families, the farm and his mates at the pub. Our first meal after the hectic moving day was a pizza I cooked in the tiny oven. I cried when I realised there were no dogs by my feet waiting patiently to be passed my pizza crusts. My emotions are never far from the surface when I am tired. James gently wiped my tears away. He made me laugh doing an impression of the man we'd hired the van off, who looked at us as if we were a bunch of teenagers going to wreck it. We opened a bottle of wine to toast our new home, and I curled up into him on the sofa like a cat nearly falling asleep.

## BASIC PIZZA DOUGH

Prep 20 minutes, plus 30 minutes for proving
Cook 10–12 minutes per pizza

Makes 4 large or 8 small pizzas

### Ingredients
400g/14oz plain flour
1 tsp caster sugar
1 tsp salt

1 x 7g packet fast-action yeast

225ml/8fl oz warm water

2 tbsp olive oil

## Method

1. Weigh out the flour in a large bowl. Make a well in the centre of the flour and add the sugar, salt and yeast. Pour in the warm water and olive oil and bring together by hand or in a mixer with a dough hook. Knead for a few minutes until you have a smooth ball.

2. Leave to rest for 30 minutes in a clean bowl or on a lightly floured surface. Heat the oven to 220°C/fan 200°C/gas 8 or fire up the pizza oven if you have one.

3. Heat a pizza stone (if using) for an hour in a hot oven. Yes, an hour . . . it needs to be really hot to get a crisp base! Or, pre-heat a baking sheet for 10 minutes.

4. Once rested, cut and shape the dough into 4 large or 8 small balls. Lightly flour the worktop and roll the dough into pizza circles, stretching out by hand to your desired thickness.

5. Top with 1–2 tbsp of passata, depending on the size of pizza, leaving an edge of about 3cm clear. Add good-quality mozzarella or a grated cheese that you like. Add any toppings you like but don't overload the pizza as this will make it soggy.

6. Slide the pizza, or more than one at a time if you have space, onto the hot pizza stone or baking sheet then into the hot oven. Bake for 10–12 minutes until the edges are starting to brown and the top is bubbling.

Note: These work really well started off in an oily frying pan – the base crisps and bubbles, ready to slide into the oven.

This morning we woke early. The sunlight was flooding into our little bedroom, the tree outside casting unfamiliar shadows on the wall. James pulled me back into bed as I tried to go and make us some breakfast, and I realised that no one is here to make us feel guilty for lying in. We made Sunday breakfast together: brown bread, scrambled eggs and smoked salmon, and a pot of coffee. I kneeled on the sofa, plate resting on my lap and fork in hand, and planned our day. There were pictures to hang and clothes to unpack. James started sorting his books out of the endless boxes scattered around the room. I measured the floor in the bathroom, scribbling the length and width on the back of my hand with a biro; I want to buy some new lino for it and made a mental note that we need a new shower curtain, rug, bin and dustpan and brush. We are in a one-bedroom basement flat in a large house in Norham Gardens. James has told me that Philip Pullman and Richard Dawkins live on the same road, and I nodded at the time, but I don't much care who they are. We are renting the flat from the college James is at. It is technically 'married quarters', but he has somehow convinced the bursar it is OK that we aren't actually married. As we sat in the sunny living room, I looked out to the sloping garden and the steps to the front driveway. 'We should get some pots and plant some herbs and lavender,' I said, but James was distracted by a box

of papers he had started to sort through. 'I'll get you some files for those,' I suggested, adding them to my list. Our priority, I told him, had to be to go and buy a fridge. So we jumped in the car and headed off to the large Currys electrical store. After the initial disappointment in discovering that their fridges took up to three weeks to be delivered, I told the store manager that we had just moved in and couldn't manage without one. He offered us an ex-display one with a tiny scratch on it, which we could take away immediately.

On the way back, with the boot open and the wind blowing my hair, we passed a warehouse selling off pine furniture. I got James to pull over. After a quick look around I saw what I wanted, and we bought a farmhouse-style table with four ladder-back chairs, which the guy said they could deliver tomorrow. I read a biography of Elizabeth David this summer, and I want life ahead of us to revolve around my kitchen table too. (That simple table is still with us today, many years later, with all its knife marks and stains.)

With the fridge finally in place, my boxes of kitchen stuff stand around in the flat with nowhere to put them for a week, so I buy a little freestanding cupboard from a charity shop and paint it before stacking my baking tins, serving dishes, trays and pans in it. James agrees (sort of) that one of his bookcases can become the glass-and-crockery cupboard and I set the ironing board up to one side for an extra worktop. We use the kitchen table for everything. Eat at it. Unload shopping onto it. Chop vegetables on it. Roll pastry on it. James studies at it.

Students are meant to eat pasta, tins of beans and lentils. But I'm not a student (and even when I was, I lived at home). I worked hard all summer before moving to Oxford and have my own money, and I think food is one of the most important things to spend it on. James worked for years before becoming a student and thinks about food like a farmer does.

'What about this?' I say as I pass him a tub of tzatziki to put in the trolley.

'OK, I'll try it,' he replies, and we wander up and down the aisles of the big Sainsbury's, with me loading up the trolley with brioche, passion fruit, figs, Greek yogurt, Parma ham and a baguette, and a load of other regular things to stock our shelves. James looks slightly anxious about my food choices, as if he's just realised he'll be spending the rest of his life with a creature from another planet.

I'll make a stir-fry tomorrow if we get some chicken, and we could have that salad I saw on TV with chorizo and potatoes. James picks up several bottles of orange juice and I say, 'You know that's full of sugar', but he replies, 'It's my house too', with a smile. We are away from the gaze and purses of our parents for the first time and for James it means unmonitored freshly squeezed orange juice – it has always been rationed by his mum.

We spend the next week in a happy blur, getting up and going out to get what we need for the day. James goes for long walks around the University Parks and down the river pathways to find ways to get out of the city. I busy myself making our flat into our home. The late-September days are warm and blue. I hang our washing on the line in the back garden and take a blanket and lie on the grass with a book. I would never do this back at the farm – everyone was always too busy for sunbathing. Everything was judged.

It is like being on holiday.

We buy a little barbecue at a DIY shop and invite some friends over for minted lamb steaks and roasted new potatoes, and I grill peach halves on the last of the heat and make vanilla bean cream to serve with them. I love making food for us. My days revolve around what we are going to eat.

### MINTED 'LAMB' STEAKS

We rarely eat 'lamb' and certainly never imported lamb. Our temperate climate and hilly ground is ideal for grazing sheep. Britain has a proud shepherding culture and history. As a family we regularly eat Herdwick hogget (a sheep of around eighteen months old) and mutton (an adult sheep) – these sheep have lived outdoors all their lives. This means the meat has been more slowly grown from herbs and grasses from the fells and is tastier than the pale, flavourless lamb found in most supermarkets that has been fattened on grain. Buying direct from a farmer or local butcher means you know where your meat has come from. Most butchers and some farmers now offer an online delivery service.

This recipe can be made with dried mint but it's much better with fresh.

**Prep 15 minutes, plus resting time**
**Cook 10–15 minutes**

Serves 4

## Ingredients

4 x 180g/6oz hogget rump steaks, brought to room temperature
2 tbsp olive oil, plus a little extra if frying

## For the mint sauce:

2 cloves of garlic
1 tbsp white wine vinegar
1 bunch fresh mint, leaves picked
a pinch of sugar (optional)
Or: 2 tsp good-quality bought mint sauce, loosened with a dash of
    olive oil and salt and pepper

## Method

1. Blitz the garlic, vinegar and mint, with a pinch of sugar if you like, then set aside. Season to taste.
2. Season the hogget generously. Cook the steaks, skin side down first, on a barbecue – the coals should be white – or in a hot pan, lightly oiled, over a high heat for 2–3 minutes on each side, basting with the mint sauce with each turn until browned all over. This will give you steaks that are pink in the middle; give the steaks a minute more on each side if they are extra-thick or if you like them a little more well done.

3. Remove the meat from the pan, baste with any leftover mint sauce and leave to rest for 10 minutes, more time if you have it, covered with foil.

These steaks can be served as they are or sliced and spread across a bed of salad leaves, sliced red onion, cucumber and tomatoes, dressed with olive oil and balsamic vinegar.

## ROASTED NEW POTATOES

**Prep 5 minutes**
**Cook 35 minutes**

Serves 4

### Ingredients
500g/1lb baby new potatoes
olive oil
sea salt and black pepper

### Method
1. Heat the oven to 200°C/fan 180°C/gas 6.
2. Bring a pan of water to the boil and cook the new potatoes for 5 minutes so that they're still firm.
3. Drain carefully and spread the potatoes onto a roasting tin, drizzle with olive oil and season with sea salt and black pepper.
4. Roast for 30 minutes in the hot oven until golden and soft.

## QUICK TZATZIKI

Makes 1 bowlful

### Ingredients
275g/10oz natural yogurt
50g/2oz piece of cucumber, halved, seeds removed, and diced
small handful fresh mint, leaves picked, very finely chopped
½ garlic clove, minced

**Method**

1. Combine the tzatziki ingredients together, seasoning well with salt and pepper, and store in the fridge until needed.

## PEACH HALVES WITH VANILLA BEAN CREAM

These are even better cooked on a barbecue – make sure the coals are white when you put your peaches on and then griddle flesh side down for 5–15 minutes, depending on the heat of your fire. You want the peaches to start caramelising in their own juices – no need for the sugar and butter.

Serves 4–8

**Ingredients**

4 peaches or nectarines, ideally ripe and ready to eat, but firm fruit will work
1 tbsp light brown soft sugar
50g/2oz unsalted butter
200ml/7fl oz double or whipping cream
1 vanilla pod or 1 tsp vanilla essence (or good-quality vanilla ice cream)

**Method**

1. Heat the oven to 200°C/fan 180°C/gas 6. Halve the fruit, remove the stones and place the fruit flesh side up in an ovenproof dish. Sprinkle with sugar and add a small knob of butter on each half.
2. Roast until juicy and soft and hot right through – about 15 minutes.
3. Meanwhile, make the vanilla cream. Whip the cream by hand or with a mixer until you have soft peaks.
4. Slice the vanilla pod open with a small sharp knife, scraping out the black seeds, then fold the seeds into the cream.
5. Serve in a bowl alongside the hot fruit. Ice cream works really well too.

I had just left college but didn't look like a typical art student; I just wore regular clothes, jeans and jumpers or printed tops. I didn't go to college to find a new identity or anything – I really liked making things and didn't know what else to do. When I started my degree I just stepped quietly out of being 'Helen' that everyone expected at home: the daughter, granddaughter and sister, the eldest, the responsible (boring) one, the hard worker and the good student. I spent time reading and writing out lots of quotes that inspired me. I made a series of black-and-white photographs of domestic objects like toasters and ironing boards placed on the fells, playing with reflections of the mountain and lake on the side of a shiny kettle. And I liked the contrast between the vast rugged scenery and the man-made lines of everyday objects.

I started sewing and made a series of strange and subversive embroideries. I was experimenting with a traditional craft, one my grandmother respected, but creating pieces that told stories of misery in the world. I had struggled initially to think like an 'artist' – because in the late 1990s 'art' was all about angst, trauma and shock value. Dead cows in formaldehyde and unmade beds with fag ends and tampons lying on the floor. I wasn't urban or alternative or angsty enough. I wanted to learn how to make carefully crafted things, and that wasn't cool. Art was meant to 'say something', and I wasn't quite sure what I wanted to say – yet. I went to see the *Sensation* exhibition at the Royal Academy. It was incredible, every piece told a story, and it had a massive impact on me. I learned to twist my work to be about darker stuff, undercurrents of violence or traumatic events that happened to regular people. I made an embroidery of people hanging from nooses from an oak tree, a mass grave covered by a carpet of flowers and a little girl standing alone on top of it all. My favourite piece that I made was a house on fire with the husband standing with his bags packed, the mother and children on the other side waving at him. I was making work about the hushed and hidden parts of everyday life. I read books about domestic violence and histories of oppression. I drove my tutors mad because I wouldn't draw out or sketch a piece before I started making it – I just started sewing and, by doing it, something emerged, the picture I had in my mind, the story I wanted to tell.

Lots of people loved my work. My favourite reaction was from the old folk who came to see my degree show: they were drawn to look closely because they understood embroidery, and then recoiled in horror at the details. I was chosen to be part of a Northern Graduates show in London and one of my pieces about a tsunami was stolen from the gallery and somehow it felt like a compliment.

I was trying to be 'Helen', a city girl with an art degree behind her and a fancy job, but I hadn't yet found this fancy job and I couldn't quite shake the home 'Helen' off, and I didn't quite know where the bold art-college girl had gone.

There was no farm work here to give James a reason not to be doing his fair share of the housework. And he did, often making meals and always doing the washing-up or putting the clothes in the machine. But I couldn't stop myself from butting in with comments like 'Don't do it like that' or 'Here, give that to me!' Looking back, I didn't give him any space. He had never done domestic tasks before and I treated him like a fool for not knowing how to fold a pile of washing. He might be able to write a complicated essay about something I didn't understand, but he couldn't boil an egg properly. The kitchen was, in my mind, my space, and I wanted to run it my way. I should have wanted him in the kitchen, but in truth I wanted him out. Making a home and making good food became my focus, a thing to hold on to when I was unsure of who I wanted to be. I was good at cooking and baking, and I loved experimenting and improving my regular dishes. I also had strong feelings about needing to sit down at the table together – never eating off our laps by the TV. And I was strict about mealtimes. We never left the flat without a proper breakfast together, then lunchtime was more flexible, but I always made us a meal every evening at 6 p.m. I didn't want my house to be somewhere chaotic where no one knew what was for supper or where I was. I wanted everything to be organised, ordered and planned, unlike the topsy-turvy way my mum had done things in my teenage years. James seemed a bit shocked that I was so bossy about all of this stuff, but he was distracted with his own things and let me take the lead. To me, everything was alright in our new world if we ate well and followed the mealtimes from our previous lives. In between trying new things, I cooked a lot of meals that were familiar to us as we learned to live in a city far away from the farms and the fields we had been brought up on. In the evenings when he wasn't reading, we'd watch *Grand Designs* and despair at the couple for taking on too much and running out of money. We agreed on which houses we liked the best when they were done.

In our tiny basement flat, I daydreamed that one day we would live in a country house like the one I'd babysat kids in a few years earlier. Velvet sofas, an Aga, wooden kitchen cabinets and stone floors. I spent too much money on magazines like *Country Living* and *Ideal Home*, admiring the photographs of roll-top baths with tongue-and-groove panelling around them, a far cry from our dingy, stained, narrow tub surrounded by mouldy tiles. Every time I talked about our future, James just fretted over how it could ever happen and how much it was all going to cost.

Late on Friday nights I'd watch every new episode of *Ally McBeal*, the hit US sitcom about a young female lawyer proving to everyone that women could rise to the top in a man's world. The Spice Girls were everywhere telling everyone what they really really wanted, and being a 'ladette' – going out, drinking yourself to oblivion and behaving like a bloke – was fashionable. It felt like being a woman who yearned to get married, start a family and make a home wasn't something I was supposed to want.

I never knew what an Indian summer was until I lived in Oxford. Autumn back home meant lighting the fire, gathering the fallen apples in and turning the clocks back. But here we were in mid-October wandering around the city without coats on, hand in hand, going out to gather brambles on an unseasonably warm afternoon. The rugby was on TV when we got back, so I had a go at making bramble jelly – it's James's favourite on hot buttered toast. After boiling the fruit up and checking the temperature with my new thermometer, I strung a jelly bag up over a bowl and watched the dark-purple liquid drip through. I ended up with about half a jar that never set. It was all a bit disheartening, so I stuck to making things I knew. Pans of soup, chicken casseroles, pork chops in cider and apples, and toad in the hole with onion gravy. It was my way of loving James. I nourished him, wrapped him up in familiarity and reassured him that he could go out and do all the strange new things that life at an Oxford college required.

## BRAMBLE JELLY

Prep 1 hour
Cook 45 minutes

Makes 3–4 jam jars (454g size); this is very dependent on how far your jelly reduces but good to have this many at the ready! Mine made just over 600ml.

### Ingredients

600g/1.3lb ripe blackberries (weigh your fruit: you want equal fruit-to-sugar ratio)
200ml/7fl oz water
600g/1.3lb caster sugar
1 lemon, juiced

### You will need

A large pan, a masher, a large ovenproof bowl, a jelly bag or nylon sieve and a piece of muslin cloth that sits over the sieve, and 3–4 sterilised jam jars (see Marmalade, p. 21)

## Method

1. Wash the brambles in a sieve or basin of cold water, then drain.
2. Place fruit in a heavy-bottomed pan with the water.
3. Cook over a medium heat with a lid on for 20 minutes to stew the fruit.
4. Reduce the heat to low and remove the lid. Mash the fruit well to release the juice, then add the sugar and lemon juice to the pan.
5. Put a large bowl in a low oven so it is hot when you need it.
6. Keep the pan of fruit cooking over the low heat to dissolve all the sugar into the juice. This takes about 10 minutes.
7. Turn the heat up and boil, rapidly stirring to prevent sticking, for 8 minutes.
8. Take the bowl from the oven and place a jelly bag or sieve lined with muslin over it.
9. Pour the hot jam mixture through the jelly bag or lined sieve carefully. You may need an extra pair of hands to help.
10. Press the hot pulp through the jelly bag or sieve quickly with the back of a spoon before the liquid has a chance to set into a jelly. If it does, you can put it into another clean pan to reheat it gently. Discard the seeds.
11. Pour the liquid into warm sterilised jars.
12. Label and store. This jelly keeps for about a month in the fridge.

James had made it through his first year in the halls of residence, with me back at home finishing my art degree, but being apart hadn't been easy. We missed each other, spending hours on the phone and writing letters to each other. That first year he was about as far outside his comfort zone as he could be. Farm boy turned student, spending whole days reading, writing essays and going to tutorials instead of mucking out barns and dosing sheep. Until recently he hadn't even been able to do basic handwriting. Before he applied, I had taught him from a children's book. Sitting together at my mum's dining-room table, he copied out the letters and tried to join them up. He knew loads of brainy stuff but barely understood basic grammar, so I worked through simple exercises with him. But he had got here and was surviving it. I believed in him.

My friends back home were getting married, had found jobs at the local NatWest bank, the town council or as teachers and nurses, and I knew they kept their finances separate from their husbands'. They told me they had joint accounts for the household bills, but after that their money was their own. They wanted financial autonomy and not to be like most of our mothers were, completely dependent upon their husbands' income. But right from the start of me and James becoming a couple we knew we wanted to pool everything, and we opened a bank account together.

After the first couple of weeks settling in, James's term started. He had to get his head down into the second year. Money was getting tight and I knew I had to find a job. We needed regular income and spending hours sewing pieces of artwork on the slim chance that I might sell one a few months later was far too risky. I had applied for a few positions through the summer since graduation, but had received no invitations to any interviews. I checked the classified section of the local paper and job sites weekly, and I immediately sent my CV out when I spotted something that I thought I could do. I received lots of letters of rejection. The honeymoon period was over.

I tried hard not to let these setbacks bother me, concentrating on thinking it was their loss, not mine, and it helped that I was away from all the questions from well-meaning busybodies back home. I put all those letters in the bin. My mum was the one who got all the questions now, friends asking her on the street, 'How is Helen getting on in Oxford?' Loaded small-town questions, where everyone is waiting for you to either fly to the moon ('Our Helen is doing really well!') or crash and burn ('I don't think it worked out, she's temping in an office.'). I don't know how she answered them. When she asked me on the phone how it was going, I always avoided the straight answer.

I needed a job. I had been a waitress since I was fourteen, and I knew I could fall back on that, but I was desperately trying to be Helen with the fancy job. That dreadful question we ask children constantly – 'What do you want to be when you grow up?' – rang through my mind every time I applied for a position. It had been asked a thousand times since I started secondary school. I had no clue of what I wanted to 'be'. My art degree had been a way of delaying that decision further. Mum and Dad thought if you had a degree you could almost walk into a career, but that idea was twenty years out of date already.

I signed up with the local temp agency, Office Angels. They sent me to work at a call centre on an industrial estate at the edge of the city. The shift started at 4 p.m. I walked in the door into a vast space full of desks, with people all wearing headsets and talking into little microphones. I was shown to an empty booth by a guy in a scruffy T-shirt, jeans and trainers, who gave me a headset and logged me in to a screen with numbers on it. I was told to phone the numbers on the list, read a script that would come on by clicking the mouse, then, if the person didn't hang up on me, click through to read further scripts. He told me that I had to hit a target of so many calls per hour and then left me to it. I cautiously called the first number with no name and said, 'Hello.' A pleasant voice of a lady answered. I apologised for interrupting her, then started to read the words in front of me. She put the phone down. I tried more numbers, and again, again and again they put the phone down on me. After what felt like forever I looked at the clock and saw it was still only 5.30. The people I was phoning were now just in from work, sorting their children out after school, doing homework or feeding their dog. Or they were cooking, and here I was, a stranger, interrupting them in their homes to try and sell them something they didn't need. I felt sick. I remembered Mum cursing calls like this at family mealtimes. I wondered what James would be having for supper alone. I thought about the ham sandwich in my bag and how it must be warm by now. I had a sudden urge to get up and take the headset off and walk out. So I did.

A few days later I started a job in an estate agent's office typing up property particulars. It was slightly better, but I longed to go out and see the actual houses. The only fresh air I got all day was walking down the street in my kitten heels to the Tesco Metro to buy my 'Meal Deal' – the same chicken salad sandwich on brown bread, salt and vinegar crisps and an apple juice. No one took a proper lunch break and the phones never stopped ringing. As I sat eating by my screen, trying not to drop crumbs down my white shirt, I imagined selling properties myself, taking couples around houses with the keys to their future dreams. But I was just a girl behind a desk, pretending to play a part. I started to avoid being caught alone in the tiny back kitchen by the manager after he complimented my knee-length brown leather boots and short tweed skirt in a creepy way one day. He was married but, unlike the other guy who never stopped rabbiting on about his family, this guy never mentioned his wife and kids. He would come into my little room and lean in too close to my screen as I clicked through the photos for the sales particulars, taking his time to choose the one he preferred for the window advert and inappropriately commenting on my perfume. A female estate agent in the team saw this happening and told me one day that he had tried his luck with her when she first started working there. I knew I couldn't stay in a windowless back office with him much longer.

SEE P.298 FOR **CREATIVE LUNCHBOX IDEAS**

## *Rabbit*

I am standing to one side of the bar, smiling and trying to laugh at the jokes, but none of them are funny. James glances across at me, rolling his eyes as if to say 'I know, they're full of shit.' I smile. I sip my drink and pretend to be having fun.

It's been a long hot day in front of the computer screen at the estate agent's. When I get home I find James lounged across the sofa, book in hand. I check the time and tell him we should be getting ready. I quickly shower and wash my hair in our tiny bathroom. I pull on my only pair of 'going out' trousers, a black skinny cigarette-style pair, and instantly feel stylish and slim, but I need a top. I hunt through my drawer and find a black one with the shop tags still attached; it has a panel of mesh across the top of the chest that feels slightly scratchy as I pull it over my head, but it will do. I rub the towel through my hair and tip my head upside down to blast it with the dryer, curling the ends under with a brush. I ask James to cut the tag and button me up at the back and he kisses my neck after fumbling with his farmer fingers trying to push the little button through the loop.

Drinks start at 7 p.m. in the senior common room. My top barely has sleeves, just a little cap over each shoulder, so I brush my arms with bronzer, trying to give some colour to my pale skin, and dab on some dark-plum lipstick that I got in a free-beauty-bag giveaway. I catch a glimpse of myself in the mirror and wipe it off immediately with a bit of toilet roll. I find some beeswax lip balm instead and pop it in my pocket. I have no idea how to apply make-up. My skincare routine is as basic as washing my face with some Dove soap on a cloth and then slathering myself in moisturiser. Sometimes I lightly comb brown mascara through my long pale lashes, but when it comes to blending eyeshadow or using a brow pencil I'm clueless. I always hurry past the stands in department stores with the flawlessly made-up sales assistants hovering around trying to sell me beauty products. Getting ready takes me less time than it takes James to have a shave and find an uncreased shirt. I squeeze the kitten heels back onto my hot, tired feet. I wear them to the office because I don't have any other smart shoes. I spritz on some Lancôme Trésor perfume that I got last Christmas and gather up my purse and my denim jacket.

James senses my nerves about going out tonight; he pulls me to him, gives me a squeeze with both hands around my waist and tells me I look gorgeous. I hurry us out the door, checking we each have a key.

I am going out because I have barely joined in with his university social events and he wants me to feel a part of it all. It is an end-of-term drinks and meal out and I don't always want to be James's invisible girlfriend.

When we get to the common room there are several people he knows clustered in a group near a piano, so we go over and join them. He has worked out a way of being, a way of behaving in this world in order to be accepted and become a part of it all. He basically hates all of this, but is polite. I tag along on his arm and smile as he introduces me. We get some drinks, and a raffle is drawn for something or other. James's name is called out and he passes me the prize. Inside the envelope is a voucher for the game stall in the Covered Market. He has won a saddle of rabbit. I laugh – only in Oxford would this be a prize for students. I have never cooked rabbit before but I suspect I am going to learn how to very soon. Back home we associated eating rabbit with being really poor and there being no other meat available – poachers' food.

SEE P.299 FOR **SIMPLE STUDENT MEALS**

No one really knows what to say to me after the initial question of what do I do. When I reply that I am a secretary in an estate agent's they move on, giving me a quick smile, and I sense pity in their reaction.

The group is getting steadily more drunk and the girls are giggling in the bathroom when I go to escape it all for a moment. One of the guys is trying to chat one of them up; apparently his family has an estate in Suffolk near where she is from. I smile and throw my head forward to puff up my hair, which is already feeling limp. I don't know why I agreed to come.

The night drags on, endless droning small talk. Our group moves moves along to the bar on the high street. It is crowded and noisy with lots of men drinking after work in suits, and women in glamorous dresses showing off their bare tanned arms and legs, their expensive handbags down on the floor next to the glass tables. They sip exotic cocktails with manicured nails. We get to the bar and James asks what I want to drink. I would normally ask for half a cider but I change my mind at the last minute and say, 'I'll have a white wine.' It comes in a giant balloon-shaped glass, so large and tall I can barely hold it comfortably and there isn't even much wine in it to keep me occupied for long. I wish I'd asked for a lemonade. I'm hot and uncomfortable in my black top and self-conscious that I look so pale. I sip the wine. My face feels red and flushed from the couple of drinks we have already had. James holds his bottle of beer and is standing a few feet away from me; he looks across at me with a knowing glance.

After a while and a few more drinks, the bar empties out and I check my watch. It is 9 p.m. and I'm starving. I ask one of the girls if we're late for our table and she tells everyone to drink up before we all head down the street to the restaurant. I have passed the door of this place a hundred or more times and never noticed it. A narrow flight of stairs that has not been cleaned in years leads us up to a door that opens into a long dark room. We huddle in around the table, decorated with tea lights and pink carnations in little bud vases. There are fourteen of us. I take a menu from a waiter who looks uninterested. Everyone is fairly drunk, so ordering the food is chaos, but they all seem to know what to ask for without looking at the menu: lamb jalfrezis, baltis, murgh makhanis, and saag aloos for the vegetarians. These are dishes I have never heard of or tasted before; back home if we ever got an Indian takeaway I would always have a chicken korma. I look over the menu when it is my turn. I don't want to appear naive, so I ask for a biryani. The food takes ages to come. I munch my way through greasy poppadoms, dipping them in mango chutney to keep me busy. I nod and listen to the chatter but have no idea who or what they are talking about. I pick my way through the biryani when it eventually arrives, having completely lost my appetite by then.

When the conversation turns to their course and the last essay they all handed in, a heated debate starts around the table. I look again at James and he can see that I am struggling. I am really tired. These students have probably slept in until lunchtime and hung around the park all afternoon, while I've been stuck in a miserable office all day. I know he can't just get up and leave – it would be rude. He is a part of this world right now and I am not. I realise I am not really part of any world.

The people around the table all seem to know exactly who they are and what they want. They exude confidence and self-esteem. Being at Oxford was always part of the plan for them. Some are off to law school, or have internships set up, or will take on a role in the family business or set up their own businesses after this. Travelling to far-away places is common for them – going to Africa or Nepal is like hopping on a bus to the next stop. Even their hobbies are extraordinary to me: rowing, skiing, climbing and surfing . . . They chase adventure, exhilaration and a life that is all about ambition and themselves. Being around these students unsettles me; it makes me question my desire to make a home and have a family. None of them sound like they will ever have children or look after elderly parents or go back home to live and work. They live with a sense of being entitled to a life that's all about themselves and will go anywhere in the world to do that.

I squeeze out of my chair. It's hot and noisy. I slip out past everyone, appearing as if I'm going to the bathroom, and whisper in James's ear, 'I'll see you back at the flat.' I walk down the stairs and into the cool of the night. I am free and relieved. These are not my people, and I don't have to be here any more. I don't even care if I don't have any 'people'. There are groups of students all over the city, spilling out of bars and hanging around the steps of the Ashmolean Museum. The heat of the day has gone and I am glad of my jacket. I take off my tight shoes and walk barefoot up the clean pavements all the way back to our cosy basement flat. I make a cup of tea and a slice of toast before curling up on the sofa under a blanket, and I have a good cry.

## Microwave Dinners

I leave the estate agent's after a few months and fall back into doing familiar cafe work. I sign up for early shifts, preferring the early mornings to the late nights. I walk the three streets between our flat and the cafe wrapped up in my scarf and hat, suede brown coat, black trousers, black shirt and comfy shoes. I am here to open up with another girl. We fire on the gas oven in the basement kitchen, daring each other to reach in further with the long matches when it doesn't catch the first time. Upstairs, I switch on the coffee machine. I lift the stools and chairs down from the tables where they were left upside down so the evening staff could mop the floor, though it's never as clean as I think it should be. The food lift fetches warm baguettes and pastries up to the bar; they have been cooked on trays from frozen, but I set them out in baskets as if they have just been prepared in some artisan bakery. I rush around making lattes and cappuccinos for men in suits queuing out the door for their morning caffeine fix. Carrying hot drinks around in paper cups and eating breakfast away from home is unusual to me – it seems like a good way to waste money. For the next couple of hours the cafe is full of middle-class mums wearing leggings and carrying yoga mats, sitting drinking cappuccinos and chatting with friends about their kids. I can't figure out how they are making enough money to be spending most of their mornings like this.

At lunchtime the 'chef' reheats mashed potato in a plastic tub in the microwave, adding margarine to it and topping it with char-grilled chicken breast that is cooked in a pan so hot it always comes out of the kitchen more burned than charred. Tubs of cooked pasta are reheated in boiling water before being mixed with scoops of squishy roasted vegetables and tomato sauce, then reheated in a pan and covered with feta cheese. The fridge is one giant Jenga stack of plastic tubs of pre-made food with dates scribbled on in coloured pen. The only things they cook fresh here are the breakfasts, but even those have rubbery scrambled eggs on the side of the plate, cooked in the microwave. The vegetarian option has some kind of strange green veggie sausage that doesn't look fit for human consumption. The owner is never here to see what the food is like; he comes at the end of the day to take the money to the bank, always pulling up right outside in his freshly polished BMW and drinking an espresso as he gathers up the heavy green money bags that we fetch him from the safe and loads them into his leather briefcase.

At the back door there is usually someone or other hanging around – the chef's boyfriend and his mates. I soon realise, on my many trips to the wheelie bins, that they are smoking weed. They don't exactly try to hide it. I am offered a joint to try one time, so, not wanting to look too square, I take it and hide it back in our flat in case the police come knocking. James lights it that night. I have a puff and feel all light-headed and dizzy. He laughs at the state of me and says maybe I am not cut out for the drugs game.

As I froth the milk at the coffee machine the next day, I daydream of opening my own bookshop cafe full of tables of interesting books and artworks on the walls. A place where the food would be cooked from fresh local ingredients and the cakes are all home-made and don't come in plastic packets from the back of a Caterite delivery van.

James often meets me after my cafe shifts and we sometimes go to the indie cinema across the road to watch arthouse films. We come out wondering what on earth they were about, and we walk into town to the Covered Market to buy something good for supper.

The farm pulls James north like a magnet. The three terms at Oxford are short so we travel back home regularly. We drive up the motorway every couple of weeks listening to Coldplay, or I read aloud. We talk about things he's working on for an essay. I ask questions, filling in the gaps of my sketchy history knowledge, and tell him that if he can explain a complicated argument to me so that I can understand it, he is winning.

It is just us in these moments, laughing together and looking after each other. But as he gets out of our car in his family's farmyard, I lose him. He goes straight to work, without even time for a cup of tea, eager to make it feel like he's hardly been away. We don't come home to chat or walk the fells or see friends. He comes home to work. In his absence, his mum has more to do outside, and he always feels guilty that he isn't there. I am a spare part at home now – I've moved out, I'm not studying any more, I don't have a job up here and coming back to stay when everyone else is working feels awkward. The farm isn't really anything to do with me, and I'm not comfortable going back to my parents' and sitting around doing nothing. I go to visit my school friend with her new baby. She is on maternity leave and seems happier than anyone else I know.

The first spring of our new city life was one we don't like to talk about much any more. Both our family farms were at the centre of the foot-and-mouth epidemic in 2001. This horrible, highly infectious disease causes blistering in the mouths and feet of cattle, pigs, sheep, deer and goats. It spread like wildfire from farm to farm across the UK, from an original outbreak in a pig unit in Essex. Our familiar countryside swiftly transformed into a killing zone. The government tried to contain and eradicate the disease by culling infected animals and other healthy animals on neighbouring farms.

James rushed back to help his mum and dad just before strict movement rules were put in place. No one was allowed on or off farms, to prevent further risk of contamination. All farms across the country were in lockdown. I stayed in Oxford, alone. I could only watch and listen to it all unfold from afar. The TV showed footage of vets wearing full biohazard suits like they were aliens from another planet, and pyres of burning animals. I phoned Mum every night after work to get updates. After a couple of weeks she told me the news we'd all feared: there was a contaminated farm nearby, and all our sheep and newborn lambs had to be slaughtered. Dad was busy helping sheep give birth, knowing that they were all destined to be culled and burned or buried soon. It was heartbreaking. The day the lorry came to load the flock, she sobbed down the phone to me. She said she could hardly bear to look the driver in the face and stood at the top of the lane as he took the animals away. The bureaucracy of it all was bewildering to Mum and until that moment I had never witnessed her crying like this.

James would hardly talk about it. He became cold and super-practical. He can make himself into a machine when he needs to. Some of the cattle on his farm were wild and had never been in a building or lorry. They struggled to round them up to be tested. After one heifer showed signs of the disease, the decision was made to call in a marksman to shoot them in the field where they grazed. The scenes he described were surreal, grim like nothing he had ever expected to see. Corpses of cows everywhere in the sunshine in the fields all around their village. He broke down when he told me that his grandfather's pedigree flock of sheep had all been killed too. I didn't know what to say.

The faces of the farmers on the nightly news were the men and women I had grown up amongst. I had served them mugs of tea and bacon butties at the auction cafe. Men that you'd never normally see showing emotion were now filmed with red blotchy faces, trying to hold back the tears for the BBC. Their life's work, their

livestock, their livelihoods wiped out. Yet here I was, 250 miles away. When I walked down the street and went to work, people around me seemed oblivious to it all – meeting friends, shopping and carrying cups of coffee back to their offices. Any talk I over-heard about foot-and-mouth seemed pointless and nonsensical, or, worse, it seemed to confirm their general feelings of mistrust in farming. I suddenly felt like a stranger in a foreign land. I was sad and lonely back at the flat. I couldn't do anything to help, I was disconnected from it all, yet it was happening in my home, to my people. My sadness felt utterly insignificant.

Later that summer I was able to make a trip home. The train passed through empty countryside, no longer populated by sheep and cattle. Our family farms were eerily quiet and lifeless. But, slowly, the farming community began to rebuild what had been lost.

I often took a longer route back to the flat after work if I was on my own, walking through the University Parks. I watched people walking together, laughing and chatting, or running, eyes straight ahead as if they knew where they were going in life. Every night while James was away I mindlessly watched TV and ate microwave dinners. I broke all my own rules about having to sit at the table to eat. Cosy under a blanket on the sofa in my pyjamas, I felt safe in the cocoon that I had made, but looking back it wasn't living, it was just existing. Endless episodes of *Friends* and *Sex and the City* were on TV, and I felt empty. I didn't have a group of friends around me all the time like the TV characters did. I knew their lives weren't real, but I slowly realised I had shut myself off from any kind of fun, friendship and support.

Oxford was a city of revolving doors. By June all of the stu-dents had vanished. Then came the tourists with cameras around their necks, forming crowds round the monuments and college entrances. People living around us kept themselves to themselves. I realised after a while that no one really knew their neighbours and there weren't any local events that helped people get to know each other. There wasn't a community that I could see. I overheard conversations in the cafe about where people had come from and how long they would be staying here. Oxford was a city that felt more like a bus stop.

I made friends with a girl who worked in the cafe; her name was Sophia, she was petite like me, but had long dark hair and a nose piercing. We worked well together, always having to hand what the other one needed. She smiled and was pleasant to the customers but if they were snooty or rude she'd mock them as the door closed behind them and that made me laugh. We often hung out after work at a nearby bar or another cafe so she could check out the guys.

One sunny September afternoon we were chatting and drinking coffee. Our shift was nearly over and it was the quiet time after the lunch rush. Someone ran into the cafe shouting at us to turn the news on. We had a tiny wall-mounted TV that the owner some-times put sport on in the evenings. All the channels were showing the same thing: scenes of a plane crashing into one of two high-rise buildings. It looked like it was in America. We stood and stared as the footage showed another plane flying straight into the other tower. I was bewildered. 'What is this? Are people still in there?' Sophia and I stood together and held on to each other's arms. The Twin Towers in New York – we couldn't take it in. I phoned James, who was up at the farm, to tell him to put the TV on. It was Sep-tember 11th and the whole world stood still in disbelief.


## Lemon Meringue Pie

The smell of brown sugar, ginger and cinnamon wafts up the stairwell from our flat as I dash to the car boot with another box of cakes. I have propped both our flat door and the main house door open and I hope they don't blow shut on me and lock me out. I have been up since 5 a.m. baking trays of chocolate brownies, lemon drizzle and caramel shortbread, and have cut it all into deep squares and wrapped it in foil parcels. The gingerbread is rising nicely and nearly ready to lift out of the oven. I am getting ready to deliver to the cafe. Two weeks ago I suggested to the cafe owner that I could bake a few cakes and we could try and sell them off the counter instead of serving the dry flapjacks and soft Rice Krispie cakes covered in cheap waxy chocolate they get from the wholesaler. I didn't say it aloud, but I felt embarrassed serving such horrible cakes. He agreed to give it a try, and thankfully didn't ask too many details about my tiny kitchen. I told him that I had an up-to-date food hygiene certificate (from a course I went on to keep Mum company as she needed it for the B&B), and we agreed some prices. We quickly sold out of everything I took, and we needed more. I am suddenly supplying a busy cafe with as many cakes as I can bake. I bake non-stop after work to keep up with demand, agreeing to twice-weekly deliveries. It is so satisfying to serve the cakes over the counter, secretly knowing that I have made them all. The reactions from the customers are wonderful. I have found myself in a world where no one seems to cook and bake at home any more. The university students all eat in their dining halls, and the street cafes, sandwich shops and restaurants are always full. In this world, old-fashioned domestic things like making good food have disappeared.

## ALMOND SLICE

Prep 25 minutes

Rest 20–30 minutes

Cook 25–30 minutes

Makes 24 slices

Ingredients

For the pastry:

225g/8oz plain flour, plus extra for dusting

100g/4oz cold butter, plus extra for greasing

2 tbsp caster sugar

2–3 tbsp cold water

### For the filling:

100g/4oz butter, softened to room temperature

100g/4oz caster sugar

2 eggs

150g/5oz self-raising flour

50g/2oz ground almonds

1 tsp almond essence or extract

7 tbsp raspberry jam

2 tbsp flaked almonds to top

### Method

1. Heat the oven to 190°C/fan 170°C/gas 5. Grease and lightly flour a Swiss roll tin (23 x 33cm).

2. Make the pastry in a food processor by whizzing up the flour, butter and sugar until resembling fine breadcrumbs. If making by hand, rub in the flour and butter with your fingertips and stir in the sugar.

3. Add the cold water and pulse until the mixture is combined into a ball. If making by hand, gently combine the water with the flour mixture with a fork, and then use your hands to bring it together into a ball.

4. Wrap or cover the ball of pastry and put in the fridge to rest for 20–30 minutes.

5. Make the filling by beating the butter and sugar well in a stand mixer or by hand. Add in the eggs and flour carefully to make a sponge mix.

6. Fold the ground almonds and almond essence into the mixture.

7. Roll the pastry out on a lightly floured surface to 3mm thick then, rolling the pastry over a rolling pin, unravel it over the tin, pressing the pastry into the corners with your fingers.

8. Spread the pastry base with jam and then spoon the filling mixture carefully over the jam, using a fork to tease it out to the sides of the tin.

9. Scatter with flaked almonds then bake in the middle of the oven for 20–25 minutes until golden.
10. Leave to cool before cutting into squares.

SEE P.307–8 FOR **LEMON DRIZZLE** AND **GINGERBREAD**

The temp agency calls and offers me a well-paid job in the John Radcliffe Hospital, a maternity leave post sorting out scans for clinics two afternoons a week. It is walking distance from our flat and I can fit it in around a few cafe shifts and my baking. The ladies that work there are lovely. As I walk up and down the corridors, I see families with their loved ones, old and young, who have undergone brain surgery, pushing them around in wheelchairs, and it is a reminder to me of how we all take our good health for granted. I know this job, however insignificant it may seem, is a vital cog in the wheel of caring for others. I call up other hospitals for the scans to be sent over on the right date and chase up any missing ones left in consulting rooms. I use my tea breaks to make lists of ingredients and ask the ladies about their favourite cakes, often taking in misshapen squares or ones that haven't risen properly, which I can't sell. I came here as the temp girl, but over the next few months I become part of their little family. After work I go shopping for extra-dark chocolate or ground almonds and hazelnuts from the health-food shop to start my baking in the evenings.

James likes my baking enterprise, a lot. He hovers around the table as I slice chocolate brownies, waiting to clean up any scraps like a hungry dog. He hopes I will slightly burn a tray of something so that I can't sell it. He doesn't care when I only offer to warm a tin of soup up for our supper. He likes to work out the costings. I buy flour, sugar and butter in industrial quantities and he often comes with me to lug it from the wholesaler or supermarket into the car and back down the stairs to the flat.

Weeks fly by in a busy whirlwind of activity. I have found my purpose and I barely watch any TV any more, unless it is an episode of Nigella or Nigel Slater cooking while I wait for the oven to heat up.

SEE P. 301 FOR **FAVOURITE COOKBOOKS**

I love making things that people enjoy. I pore over recipe books and challenge myself to make giant lemon meringue pies; they turn out exquisitely. I perfect my short-crust pastry and fill them deep with golden-yellow lemon curd and pile them high with soft mallowy meringue. I make flapjacks with cranberries, pecans and caramel swirled on top, and rocky road bursting with cherries and marshmallows. Chocolate fudge cakes with cream and dark chocolate icing, so rich you feel wired after eating a slice. And crumbly and gooey date and oat squares, sizzling up the filling in a pan, then adding lemon juice to cut through the sweetness of the dates. I try to perfect old recipes from home. I work on my coffee cake until it is so good that people sneak back to the cafe for a second slice after their friends have gone down the road.

## LEMON MERINGUE PIE

This recipe is for a 20cm fluted pie tin. If your pie tin is bigger, or you're not confident you can roll the pastry thinly, double the quantities and wrap and freeze any leftover pastry.

Prep 40 minutes
Rest 20–30 minutes
Cook 45 minutes

Serves 6–8

## Ingredients

### For the shortcrust pastry:

110g/4oz plain flour, plus extra for dusting

25g/1oz icing sugar

55g/2oz cold unsalted butter

1 egg yolk

2 tbsp cold water

Or: use 200g ready-made shortcrust pastry

### For the lemon curd filling:

juice of 2 lemons

2 egg yolks

140ml/¼ pint water

25g/1oz cornflour

75g/3oz caster sugar

### For the meringue:

2 egg whites

75g/3oz caster sugar

## Method

### Pastry:

1. Heat the oven to 180°C/fan 160°C/gas 4.
2. Whizz the flour, icing sugar and butter together in a food processor or by hand to a crumbly, sand-like texture.
3. Add the egg yolk and water, and mix again to form a dough. Bring together into a ball on a lightly floured surface if it hasn't come together already. Wrap the pastry and chill in the fridge for 20–30 minutes.
4. Roll the pastry out to 2–3mm thick between two sheets of greaseproof paper to prevent it sticking to your surface, and line the dish, gently pressing it into the corners of the tin with your knuckles.
5. Cover with a sheet of greaseproof paper and tip in enough baking beans or any dried lentils, rice or dried peas to reach the top of the pastry case.
6. Bake the pastry case in the middle of the oven for 10 minutes.

Remove the paper and beans and cook for a further 5 minutes until golden and crisp.

## Lemon curd filling:

1. In a pan whisk the lemon juice, egg yolks, water, cornflour and sugar. Warm over a low–medium heat until it starts to thicken, then increase the heat slightly and cook for a minute more, whisking, until you have a thick glossy curd that coats the back of a wooden spoon.
2. Pour the curd into the baked and cooled shortcrust pastry base and smooth it with a spatula. The pastry and the curd can be prepared and assembled a day ahead and kept in the fridge.

## Meringue:

1. With the oven set to 180°C/fan 160°C/gas 4 again, whip the egg whites to stiff peaks in a clean dry bowl – 3–4 minutes in a stand mixer or with an electric whisk.
2. Add the sugar bit by bit until all combined and the sugar has dissolved. Check this by rubbing a little of the meringue between thumb and index finger; if it feels granular whip it some more.
3. Gently spoon large clouds of the meringue on top of the pie, spreading out to the edges and adding peaks with a fork or the back of a spoon. Bake for 15 minutes until golden.

# CRUMBLY DATE SQUARES

Prep 15 minutes
Cook 30 minutes

Ingredients
For the filling:
200g/7oz dried dates, chopped and pitted
50g/2oz soft light or dark brown sugar
1 tsp vanilla extract
1 lemon, zested
150ml/¼ pint boiling water

**For the oat base and crumble:**

- 100g/4oz butter
- 85g/3oz soft light or dark brown sugar
- 2 tbsp golden syrup
- 200g/7oz rolled oats
- 1 tsp ground cinnamon
- 85g/3oz plain flour

**Method**

1. Heat the oven to 180°C/fan 160°C/gas 4. Make the date filling. Put the chopped dates in a pan with the sugar, vanilla, lemon zest and boiling water. Stir over a low heat until they have absorbed most of the liquid. Allow to cool.
2. Grease and line a 20 x 20cm brownie tin, then make the oat base and crumble. Melt the butter, sugar and golden syrup in a pan. Stir in the oats and cinnamon and fold in the flour.
3. Press half of the oat mixture into the lined tin.
4. Spread with the cooled date mixture.
5. Crumble the rest of the oat mix on top and press down gently.
6. Bake for 20–30 minutes until golden. Allow to cool then cut into squares.

Alternatives to the date filling: rhubarb with orange zest/juice and a little sugar to sweeten; brambles cooked into a jam with orange/lemon juice and sugar, with a few left whole; raspberries and apples cooked in a pan with a little sugar and water.

I soon start making enough money to cut the cafe shifts down to concentrate on baking. I am freer in the days to lounge by the river with James for a picnic on a sunny afternoon and then work late into the night while he is studying. We put music on in the flat and dive deep into our worlds of baking, reading and writing.

My aunt drives my farm grandma down to see us. They stay in a posh hotel in Oxford. Ever since we left, she has been itching to come and visit, my every move no doubt turned into a proud boast at the shops – 'Our Helen is doing really well in Oxford.' She has written me lots of cards and letters and when she heard I had started baking she copied out her favourite recipes for cherry Genoa cake and ginger biscuits.

## GRANNY ANNIE'S GINGER BISCUITS

Prep 15 minutes

Cook 15 minutes

Makes 22

### Ingredients

110g/4oz unsalted butter

1 tbsp golden syrup

170g/6oz caster or soft light brown sugar

1 large egg, beaten

340g/12oz self-raising flour

2 tsp ground ginger

1 tsp bicarbonate of soda

### Method

1. Heat the oven to 180°C/fan 160°C/gas 4. Line a baking sheet with greaseproof paper.
2. Beat the butter, syrup and sugar together, using a stand mixer or by hand.
3. Add the egg and combine.
4. Add the flour, ginger and bicarbonate of soda to form a dough.
5. Scoop into balls with a spoon or your hands, flattening slightly, and bake on the prepared tray for 15 minutes until golden.
6. Transfer with a palette knife to a cooling rack when they become crunchy.

SEE P.309 FOR **COFFEE CAKE**, P.311 FOR **DROP SCONES** AND P.312 FOR **SCONES**

It is a warm day when they arrive, and despite the two flights of stairs being difficult for her bad knee, Grandma hobbles across the room and sits down on the green sofa, a hand-me-down from her friend Audrey. I open the window for some more air, and she tells me how lovely I have made it all. They enjoy a cup of tea after their long journey. I've baked them sultana scones to be served with jam and cream like a proper posh afternoon tea, but my aunt just has a biscuit because she reminds me that she doesn't like dried fruit in anything. James laughs and jokes with Grandma. She likes his cheeky comments as he asks her about her past. She tells him stories of riding on the back of his great-uncle Jack's motorbike before she got married – 'He was fast in more ways than one,' she chuckles. I watch Grandma's face; she is taking it all in as if she is storing every detail of our new lives in her mind. The books on the shelves, the plates, cups and bowls, pictures on the wall, the flowers in the pots outside and my tiny kitchen at the back of the room. As she leaves, she hugs me and says she is happy now that she can picture me here. She is nearly ninety and I know she won't come again.

The hospital job came to a natural end as it was a maternity leave post. The ladies got together to wish me well on my last day, and we shared a cake and a quick cup of tea.

I spotted an advert in the paper for a seasonal job managing orders for free-range geese and turkeys on a local farm. I knew I was the ideal candidate – a farmer's daughter who had sold turkeys from home all her life. They had seen a few people, but I got the job – I took a cake with me to the interview.

I liked being back on a farm. I had a scruffy farm office desk to work at, with an old black Labrador that lay at my feet and a feed merchant's calendar on the wall to tick off the days until Christmas. It felt like home. The window overlooked fields full of traditional bronze turkeys that pecked in the grass and gobbled at every little insect or worm they spotted. On the other side of the lane there was a flock of pristine white geese with bright-orange beaks grazing and honking every time a dog or car went past. Instead of taking names down in a notebook like I had done for Dad, I had to learn a more complicated computerised ordering system, and work quickly as the phone was ringing constantly. I used my best phone voice, trying not to sound too northern. Back home, Dad and I would weigh the birds and they went on a 'first come, first served' basis – most people got the size they wanted, but if a couple turned up late on Christmas Eve there might only be a giant turkey left. Dad would offer to cut it in half for them or take the legs off there and then, and because everyone who bought from us was a family friend they were always very obliging. But in Oxford, these were very different customers, and the business was on a much bigger scale.

People who shopped at the farm shop had fancy ideas about themselves. They wanted to discuss with me what weight, in kilograms not pounds like back home, of turkey or goose would feed seventeen people. They would chat for ages about how long to cook it for. They would not have been happy turning up to find all the medium-sized birds gone, and some farmer offering to cut bits off a gigantic leftover bird to get it stuffed in their oven. As Christmas approached, I realised how posh the whole enterprise was.

The farm staff set up a white marquee near the front of the little farm shop as a collection point, and we set tables inside it, stocked full of their own farm-grown potatoes, carrots, parsnips and Brussels sprouts. James and my brother Stuart – who was visiting us for a few days – got roped in behind the counter on the final weekend to work as kind of pretend-butchers wearing aprons and Santa hats. It was a crazy-busy three days, but we loved the energy of selling the farm produce direct to the customers. We all drove back home together very late on Christmas Eve that year, with the radio blasting out the car and pockets full of cash from all the tips. Dad couldn't believe how much per pound they were charging for turkeys in Oxford when we told him; he thought it was like a goldmine. We had supposedly left farming behind, and yet it kept pulling us back in.

The owner of the little farm shop next to the turkey farm was called Gill, a short stout woman who wore pearls and neat shirts with the collars standing up. She always had an apron tied around her waist and peered over her glasses at me when she fetched me a coffee and a bun. 'Everything OK, my dear?' she'd ask, and I'd smile: 'Yes, all good, thanks' – lying to her because I didn't really know what I was doing when I first started the job. Gill sold the farm produce, but her passion was cheese. She had a counter full of speciality cheeses and she'd won a British Cheese Award for her shop. The smell of Stinking Bishop and Stilton turns my stomach every time I think about that counter. I have never been a cheese lover (after being sick from eating it as a child) but I quickly realised in a cheese heaven like this I had to keep quiet about my aversion to the stuff.

Gill also catered for all sorts of events locally – funerals, weddings and parties – and I offered to help her.

I sensed she was suspicious about my offering at first, probably wondering how useful I could be in the kitchen. I had been hired to be the farm secretary, after all. But after working for her a couple of times I proved myself, and soon I became a core part of her team. We were hired by people with fancy houses to take over their kitchens and set up delicious lunches of roast pork stuffed with prunes and apricots, and whole salmon baked with dill and lemon and herby new potatoes. We made platters of sandwiches and cakes for garden parties for tennis players who had swimming pools, and we prepared afternoon teas for racehorse trainers in marquees. Gill's speciality was to assemble huge cheese wedding cakes decorated with flowers and fruit, something no Cumbrian had ever seen at that time. Fruitcake was the only kind of wedding cake we were likely to have. I got to see how the other half lived, how they sipped champagne on terraces, and employed caterers like us to scurry around like mice making it all look and taste wonderful and then disappear, leaving it all immaculate, as if we hadn't been there at all.

Gill was strict but fair to work for, though she wouldn't let me anywhere near her cheese-loving customers once she discovered I wasn't keen on it; how could I know what to recommend if I hated the stuff, she said. So I made lunches for the pensioners who regularly came along to her tiny cafe. I offered to bake cakes for her to sell in the shop, and after she had tasted my coffee cake, I was in. Soon after that she let me use her kitchen on a Sunday as she had a much bigger oven than I did back at the flat. I loved working away in there with my boxes of ingredients and cake tins and always left the place spick and span, loading up the warm cakes for the cafe in boxes in the back of our VW Golf.

We had enough money to get by and quite a lot of free time, with very few obligations to other people. I had found a way of living in a city that felt true to my past and the things I cared about. We ate out regularly, spending the cake money at the weekends, exploring the Cotswolds and trying out food in the gastropubs opening up everywhere. Pizza Florentine with spinach and an egg cracked on top or a salad of torn chicken with slices of crisp Granny Smith apples topped with crumbled blue cheese were James's favourites. I would often order posher versions of what I had grown up on: tasty shepherd's pie made with a sweet potato mash instead of lumpy potatoes, or a rich steak and ale pie with crisp flaky pastry and buttered greens. Then I would attempt to recreate our favourite meals from their menus.

## Santiago

The old people spin and whirl around each other on the dance floor in the ballroom of the hotel to the music of the jazz band on stage. We are seated at big round tables lit with candles and decorated with flowers as if it is a wedding reception. I chat to a young woman from Peru with long dark shiny hair and a face that could launch a thousand ships. She compliments me on the colour of my red hair and admires the pale skin of my arms. She makes me blush – no one has ever mentioned my appearance like that before, apart from James. She is here on a similar scholarship to James and her English is impeccable.

James won an essay competition earlier this year; his tutor had suggested he enter, and he wrote it at my mum's dining-room table over a few evenings of the Easter holidays. 'Chile!' he laughed down the phone as he told me. 'I've won that essay thing and they're flying me to Chile to speak at their conference.' It couldn't have been further from our wildest dreams at that point, to be flown across the world at someone else's expense.

'If you're going, I'm coming,' I said, drying my hands on my apron in the kitchen of the cafe I was working in. I spent the next week or two checking with the organisers that I could join the 'spouse tours' that they'd suggested in a few email exchanges. I booked the same flight as James had a ticket for, and requested our seats be together.

When we arrive in Santiago it is mid-November. The warm dry air makes me feel light and free – back home it was cold, and the days were dark and short. We notice a driver holding a placard with James's name on it as we walk through the airport arrivals hall and feel like royalty. We are driven to a high-rise hotel in a smart black car, watching the city taxis screech by, leaving people wrestling luggage onto buses behind us. The hotel is mostly glass with a waterfall in the lobby. We check in and I feel mildly self-conscious as I realise the clothes that I've brought are mostly easy travelling clothes; I have almost nothing to wear that could be described as smart, and no stylish handbag to fit in with the posh businesswomen and golf widows I glance at in the foyer. We drop our bags in the room and go off exploring before the conference starts the next day. We walk to the market and eat empanadas and sopapillas (pumpkin fritters) and walk past stall after stall laden with every kind of fresh fish and vegetable imaginable.

We have been given a full itinerary for the week. James has lectures and discussions to attend and will present his paper in the last session. Every morning I join the partners (mostly elderly wives) of the delegates on organised bus trips, leaving from the front of the hotel and driving out to markets and local beauty spots with lunches included and plenty of opportunities for shopping. The ladies take me under their wing; I am the youngest here, whereas they have been to these kinds of things before and a few of them are friends already. They are mostly wealthy Americans dressed in bright patterned blouses with linen trousers and trainers and they instantly put me at ease. Despite feeling a bit of a fraud as I'm not officially a 'spouse', I chat to them about where I'm from – 'the English Lake District' – and they're fascinated, and impressed that James won the competition that got us here.

# LAMB EMPANADAS

Prep 25 minutes

Cook 1 hour 15 minutes

Makes 6 large pasties

## Ingredients

1 onion, finely chopped

50g/2oz mixed veg: ½ red or green bell pepper, chopped, a few
mushrooms, green beans, ½ courgette or a medium carrot, diced

200g/7oz lamb mince

¼ tsp red chilli powder

½ tsp salt

1 tsp cumin seeds

¼ tsp ground cumin

¼ tsp ground coriander

¼ tsp ground turmeric

2 spring onions or ½ onion, finely chopped

50g/2oz frozen peas

320g/11oz ready-rolled shortcrust or puff pastry

1 egg, beaten, or a little milk in a cup to use as a glaze

raita or yogurt and/or pickle or mango chutney to serve

## Method

1. Heat the oven to 200°C/fan 180°C/gas 6. Heat a little oil in a frying pan
   and cook the onion and diced vegetables, reserving the spring onion,
   for 10 minutes until soft. Scoop them out into a bowl and set to one side.

2. Using the same pan, fry the mince and salt over a medium heat,
   breaking it up until it starts to brown. Add the onion and vegetables
   back in and stir in the spices.

3. Cover the pan with a lid and cook gently over a low heat for 30 minutes
   to make sure the mince is fully cooked and the vegetables are tender.

4. Once cooked, leave to cool completely and stir through the spring
   onion and peas.

5. Lay out the shortcrust or puff pastry and cut into 6 squares (roughly 15cm x 15cm). Fill the squares on one corner with the cold lamb filling, making sure to leave a border clear. Brush the edges with egg or milk and fold over the pastry to make a triangle, pressing and sealing it into a pasty, crimping the edges if you like. If you're finding it fiddly, remove a little of the filling.
6. Prick the top with a fork or knife and place on a lined baking sheet, brush with the egg or milk and bake in the oven for 15–20 minutes until golden brown.
7. Serve with raita or yogurt, pickle and mango chutney.

I am drying my hair upside down in the hotel bathroom and I notice in the mirror that my face is red from the sun, so I apply a little foundation from a tester sample in a tiny make-up bag that I have brought. I lift the wooden hanger from the wardrobe with my dress for tonight. It is a full-length midnight-blue velvet dress with fine beaded embroidered flowers running up from the waist across the front, with a high, straight neckline, two thin straps leaving my arms bare and a long slit up the back from the ground. It feels luxurious to touch and I stared at it for the past few weeks hung in our bedroom before packing it carefully in a cotton bag in my suitcase ready for tonight. Mum and I had been to the boutique in town looking for a dress suitable for a black-tie dinner, and we found this one almost hidden between the frilly, puffy ballgowns in coral and turquoise. I'd been about to give up hope, but I pulled this dress out from the rail and knew it was the one.

I pull the dress over my head carefully and stand to check myself in the mirror on the wall. From the hotel window I can see the jagged snow-capped peaks of the Andes. I look taller and slimmer and feel more elegant than I perhaps ever have in my whole life. I slip on the silver sandals I have borrowed from my sister for tonight and pick up the glittery purse I bought in Accessorize that matches the dress. I am ready.

James is standing in his brand-new stiff white shirt, fiddling with the bow tie that he needs to wear. I clip it onto his shirt and straighten it. He pulls on the dinner jacket we have borrowed from his cousin's seventy-year-old husband; I tell him he looks like a penguin and we laugh. We have never dressed up like this before. He looks at me and tells me I look gorgeous. He puts the room key in his pocket, turns off the lights and we head down the corridor together. I check myself in the lift mirror to make sure this isn't a dream, and quickly apply more lip gloss.

We eat, drink and chat to the new friends we've made, swap addresses and faithfully promise to stay in touch. We watch the dancing, and I can feel the love these older couples have for each other as they twirl the night away. We have never danced in a ball-room and have no idea how to waltz or tango, so we sneak away outside for a walk and some fresh air.

It is a clear, calm night, and I am slightly tipsy. I clutch onto James's arm as we walk. The roses are in full bloom and their heady scent draws us into the park. James pulls off from holding my hand, turns in front of me to stop me walking and looks at me as he gets down on one knee. As he holds both my hands in his, he asks if I will marry him. We are standing in a rose garden under the silhouette of the Andes Mountains. Tears, and laughter, and I am nodding: yes, yes, yes. We kiss under a sky full of stars.

# AFTERNOON

'Come on, Floss, we'll go and tag some lambs.' She nudges my hand with her nose as I pull my boots on. I grab the medicines we'll need from the fridge. James has brought the other dogs down from the shed, including Bess, who is ready for a break from her pups. 'See you in a little while,' I call to Tom as we drive off. 'Be good for Granny!' Mum waves to me from the flower bed with her rubber gloves on as she wrestles with a giant dock weed.

Down at the sheep pens I get the box out and start sorting a few coloured ear tags. The field across the road has ten ewes and lambs in it; they are nearly a month old and are fit and strong. I walk across and open the gate for James to drive the quad bike around the field. He sends Bess and Tosh around the sheep and I stand on the road with Floss, ready to stop any oncoming traffic and turn them into the pens. 'Away!' he calls to both dogs, and then: 'Fetch on.' The little flock runs out the gate and past me, bleating noisily to each other, lambs and mothers mixed up. Bess runs ahead to turn them into the yard on James's command. We give them ten minutes to mother up. The sheep have individual sounds and smells, and once reunited the lambs suckle for reassurance. Floss is in exactly the right spot: she has cornered two ewes with their lambs next to the first gate of the pens. I quietly walk behind them, clapping my hands a little to chase them in. I give Floss a well-deserved 'Good girl' pat. One of the ewes has twins, and the other a single lamb. I scan the first ewe's ear tag with an electronic device that reads her microchip number. I look through my five different lists of sheep numbers to find her.

'She went to The Jedi,' James calls out, which immediately helps narrow my search down. The Jedi is one of our five tups (male rams) that we put with separate flocks of ewes to be mated last November.

'I've got her here,' I say. She is on the right list, and I tick the number off.

Knowing the father of the lamb tells us which colour tag to put in its ear, so that for the rest of its life we can tell its pedigree. James can remember which tup he put every ewe to, hundreds of them by sight, and has a photographic memory of them. We specialise in selling prime breeding tups in the autumn sales to other shepherds across the Lake District and beyond. Any male lambs that don't make the grade for breeding as they grow are castrated and sold for fattening in the autumn. I bend the white-coloured plastic ear tag into the handheld tagger and pass it to James. He clips it into the lamb's ear, and I pass him the corresponding yellow electronic tag for the other ear. The lamb is startled with its new ear piercings and kicks and makes a loud bleat to its mother. I fill and pass James a small syringe filled with the vaccination medicine.

'A tup lamb,' he calls to me. 'Its mother was a show ewe.' I write some notes down in the record book and he gives it a score out of ten. 'This one is a seven,' he says, 'it's got potential.' We keep handwritten records of all the sheep. I write down how the ewe is faring after lambing time as he checks her teeth and udder. She is a fell ewe, and this is her third lambing. 'She's a nine,' he calls, and I write that down too. It's our own system, designed to guide us when we come to deciding whether to keep or sell any sheep. It helps us have a good overview of the health of the flock too.

James pulls a little dirty wool off her backside, keeping her clean for the lamb to suckle. I pass him a stick covered in blue-coloured smit paint to dab on the top of her back, and then the red, to make the mark that tells all the other shepherds on the fell that she is our sheep. I pull the cord that opens the gate for them both to go into the next pen through the footbath. James goes to set the lamb down and it leaps out of his arms, splashing through the water as it runs off.

James lets the next ewe in with twins. I pick one of the lambs up. Its body is warm, its black fleece thick and soft beneath my fingers. Its legs are sturdy with clean little hooves. 'A gimmer,' I say – a female lamb. Her pink tongue trembles as she calls to her mother and I cuddle her into me, reassuring her, and tell her she'll be back with her mum in a minute.

Every ewe and lamb gets the same treatment – scanning, tagging, writing records, injecting and marking. The sun is shining, and the sky is blue, and it's a good day because we haven't fallen out (yet). It takes us just over an hour to finish this batch. Floss is dozing on the back of the quad bike, Tosh is charging around, Bess is lying still under the gate.

When we are done, I open the gate onto the road and walk ahead a few yards to let the sheep into the next field. But as I look back, I start to run because Tosh has rounded them all up and has already got them out onto the road behind me. 'Lie down, Tosh!' James shouts at him, but he is young and keen to work and he wags his tail proudly and weaves quickly from side to side behind the sheep and lambs. I struggle to push open the gate in time, lifting it over the clumps of long grass. The flock races past me, the ewes pushing each other into the new field. They know where they are going, but one lamb runs too far past me, and Tosh runs round it and turns it back to the gateway, and it runs in. Soon their heads are down, grazing. They are happy, their lambs beside them. I close the gate. It will only be a couple of weeks until some of these sheep and lambs go up onto the high fells for summer grazing, and then we will let the meadows rest, to grow grass to make into hay for the winter.

I climb back onto the quad bike. We need to go and check the new calf. The dogs nuzzle into my hair as we set off, pushing past

HELEN REBANKS 125

each other to get the best position to see ahead over my shoulders. Tosh makes me laugh as he rides with both front paws on my back. James drives along the straight road that is shadowed from the high afternoon sun by the line of oak trees that give the field its name. He slows before the bridge, and then stops and stands to let me off. I open the gate into the field. The herd is now sheltering under a few trees near the fence, so we drive across the field, bumping over clumps of rushes and the gullies left by the last heavy rainfall. A heron flaps up awkwardly into the sky just ahead of us; we've disturbed it from hunting for frogs in the pools. Heidi is resting and her calf is tucked in close to her, almost out of sight. All is well and we don't disturb them.

I pull up next to Isaac's school before the other parents arrive. The primary school is nestled into the valley bottom, dwarfed by mountains. It has bright-blue-painted gutters, a little metal gate and a welcoming front door, and is built from the local quarried slate stone. Herdwick sheep graze the fields around the school. It is very unusual for me to be early.

'I need a wee!' Tom shouts from the back of the car.

'OK, undo your seatbelt. I'll get you out and we'll cross the road. You can do it in the field, I'm sure Graham won't mind.' Graham fetches his daughter along to school most mornings in his farm truck with a bag of sheep feed in the back.

I fasten Tom safely back in his car seat, leaving him watching me collect Isaac from a few feet away with the window open. On the roadside I curse the holiday traffic. Cars whizz past us on their way to Kirkstone Pass, where they will wind over the little roads through the fells to the busier tourist spots of Ambleside and Windermere. Several other parents and grandparents arrive. We all smile and talk about the weather but mostly keep ourselves to ourselves.

I hear Isaac and his friends coming before I see them. His teacher reaches over the wooden gate to the latch and unclips it.

She apologises for being a few minutes late because they wanted to finish a piece of music. The kids jostle out and she checks to see who is here to pick them up. She sees me and lets Isaac go. He opens the car door but can't get in properly for boxes of blue, white and brown eggs. I lift a couple off the front seat, and on impulse get rid of them by giving them to his teacher, who thanks me. She says quietly that his reading aloud today was brilliant, and that I should ask him to read the passage to me later. She says he had the whole class in stitches with the characters' voices. I smile. He hates getting any praise in public, so I don't make a thing of it. We wave to her and his friends as we drive off.

Isaac looks tired. I ask him what's up and he grumbles about not being allowed to play the game they started at lunchtime with sticks as guns and pinecones for grenades. He lights up as he tells me all the hiding places they made: 'I found the best stick for a sniper rifle.' I ask him if he knows why his teacher has banned war games. 'Well, yeah . . . I know war is really bad,' he says in a serious but disappointed tone.

'I want to go to the ice cream shop, but not for ice cream,' Tom interrupts. I need some bread, so I say we will call in. We pull over and park on the road at the front of the village shop, a mass of rucksacks, umbrellas, plastic balls and waterproof coats hanging under the awning. Tom pleads with me to let him go in. Sometimes, when I am in a rush, I make him wait in the car. Today I let him climb out between the tourists queuing for ice creams. He shoots off into the shop. It is an Aladdin's cave of groceries, gifts and outdoor gear. The shopkeeper, Rob, says, 'Hiya, Tom', and Tom looks up and grins. There is an old lady waiting to buy some carrots and a pie, and a crowd of teenagers probably on some Outward Bound adventure holding cans of fizzy drinks, crisps and bars of chocolate. Tom has picked up a packet of my favourite biscuits. He offers me them before heading around the corner to the sweets and toys. He thinks that if he softens me up he will get what he wants.

And, frankly, he might. I'm ready for a cuppa and a biscuit. He picks up a shiny, soft, sand-filled turtle. I say, 'You can't have sweets *and* a toy . . . You'll have to choose: is it chocolate or the turtle?' He looks longingly at the sweet aisle and then hugs the turtle close to his face. 'I love this turtle, I want to call him Steve!' We take the turtle, a bag of frozen peas, a carton of milk, the biscuits and a loaf of bread to the till. Rob charges him less than the full price for his toy and I let Tom pay with a few coins I have in my pocket from the egg sales yesterday.

# 3

## Roast Beef and Yorkshire Pudding

Water gushes down the canvas flap hanging over the doorway as I pull it aside. I step into the empty marquee and it feels gloomy and miserable. It looks like a beer tent at an agricultural show. The grass squelches under my dirty wellies. The white plastic panels are stained and rain trickles down the transparent panes at the front, obscuring the view. I shiver with cold. I am wearing an old brown fleece jacket of my mum's. It is 7 a.m., the day before my wedding. It all looks terrible and I wonder what I have done.

Soon there are people everywhere, bustling around doing stuff. My family, James's family, friends, and Mum and Dad's neighbours, all asking questions at the same time.

'Where shall I put these boxes?'

'Which corner shall we make ready for the old folk?'

'How do you want to decorate the entrance?'

'When are the caterers coming?'

The marquee company people arrive and start to put lining fabric inside it. Ivory-and-green swags cover the plastic, masking the grubbiness. A lorry with tables and chairs arrives, an hour late, and Dad and Stuart unload them. Dad is grumbling that the driver has churned the field up with his tracks and that it'll be a miracle if he gets out without getting stuck. Mum is tutting and flapping about the state of everything. It hasn't stopped raining for a week.

My grandma's words play on repeat in my mind as I work, carrying boxes from the house. 'There's neither style nor comfort in a tent,' she said when we announced we would be holding the wedding reception in a marquee in a field. I know Grandma would rather it all be at her preferred hotel like a 'proper' farming wedding: roast beef and Yorkshire pudding, with silver cutlery, and waiters and waitresses pretending to treat all the guests like royalty, formal photos on the steps on the lawn by the roses, and then another wedding next weekend with the same routine. I stood against all that, I wanted to do my own kind of wedding, and it seemed like a good idea at the time. But now I know that if it doesn't stop raining we will all be sitting around in thick coats over fancy dresses, with the women's heels sinking into the sodden grass on the walk from their cars to the marquee. I am determined to prove Grandma wrong, so I whisper a hundred little prayers for the sun to come out.

## ROAST BEEF

Sourcing your beef is the most important part of this meal – buy the best you can afford and look out for special offers of beef boxes direct from farmers. Joints to choose from are rib (on the bone or boned and rolled), sirloin, top rump and fillet. (Silverside and brisket are best braised for longer, slower cooking.) Be aware that meat may cook less evenly if bone-in; make sure you allow for good resting times and if in doubt use a meat thermometer.

**Prep 15 minutes**
**Cook 2 hours** (depending on the size of the joint)

Serves 6–8

### Ingredients

1.8kg/4lb roasting joint of beef (as above), brought to room
   temperature
2 onions, sliced
500ml/17.5fl oz fresh stock or water from cooking vegetables
1 tsp redcurrant jelly or bramble jelly

100ml/3.5fl oz red wine

2 tsp plain flour mixed with 1 tbsp water

## Method

1. Heat the oven to 200°C/fan 180°C/gas 6. Season the meat. Slice up one or two onions and place in the middle of the roasting tin.

2. Place the meat joint on top of the onions and put in the oven, basting it regularly with the juices as it roasts.

3. Roast the beef according to its weight. Work on a rough calculation of 15 minutes per 450g/lb, plus 15 minutes to heat through, and then factor in 20–30 minutes' resting time. For a meal at 6 p.m. I would start cooking a 1.8kg/4lb piece of beef at 3.45 p.m. and give it an hour and a quarter in the oven – testing the meat with a skewer towards the end of cooking to see if the juices are starting to run clear. We like to eat our beef pink so I would take it out before they are totally clear and leave it to rest under foil. You can also use a meat thermometer for greater accuracy – for medium rare the thermometer should read 52–54°C when inserted into the centre of the joint straight out of the oven, for medium 58–60°C, and for well done 65–68°C.

4. Remove the meat from the roasting tin, scrape the onion and juices into a small pan and add some stock or reserved vegetable cooking water (I always use carrot cooking water for my gravy).

5. Add the redcurrant or bramble jelly and the red wine, and whisk in the flour/water paste. Stir over a medium heat until thickened. Strain through a sieve into a warm gravy jug.

## YORKSHIRE PUDDINGS

**Prep 5 minutes / Rest 30 minutes**
**Cook 20 minutes**

Makes 12

## Ingredients

100g/4oz plain flour

½ tsp fine sea salt

2 eggs

280ml/½ pint milk

100g/4oz lard for the pan

**You will need**

A deep 12-hole muffin pan

**Method**

1. Heat the oven to 220°C/fan 200°C/gas 8 or increase the oven temperature after you've taken the beef out. Sift the flour and salt into a mixing bowl.
2. Make a well in the centre of the flour and crack the eggs in.
3. With a whisk, start to combine the eggs with the flour, adding the milk a little at a time to make a smooth batter. Pour it into a jug and leave to stand in the fridge for 30 minutes.
4. Scrape 1–3 tsp of lard into each pan of a deep 12-hole muffin tin, and put in the hot oven for 2–3 minutes.
5. Take the hot tray out of the oven, being mindful not to spill the fat, and carefully fill each muffin pan half-full with the cold batter. It should sizzle and bubble as it hits the oil.
6. Cook near the top of the oven for 10–15 minutes until risen, golden and crisp but not dried out.

Every spare evening and weekend for the past few months I have been planning, writing lists, ordering things and making all the wedding stationery by hand. Order of Service cards sit carefully boxed up in the house, little sprigs of lavender glued to the front, with the hymns that I typed up, ready to take to the church later. Another delivery arrives: a neighbour who runs a plant nursery has brought several pots of trailing ivy to decorate the poles of the marquee. 'Get a ladder!' someone shouts, and I say, 'Hang on, I want to wind some roses into it all first.'

Mum chats to her neighbour about all the work there is still to do, and I catch the sound of nerves in her voice. Stuart is sent to find more extension cables and then the caterers arrive, filling the side tent with a mobile oven and more tables. The band are here; they carry their speakers in and after a while music blasts out of them as they test the sound. Mum starts to dance with a brush in her hand and shouts to the band, 'You can leave it on!' My sister Alison is stringing storm lanterns onto the washing line in her rain-coat. The floor is getting muddier and muddier, but everyone seems to be enjoying doing something completely different from their normal daily routine. I revel in the hustle and bustle of it all. There is something lovely about all this old-fashioned working together.

James is strimming grass along the lane in the rain, deliberately keeping out of the way. He always said that he could live with-out getting married, that it was just a piece of paper and that he loves me, so why spend all the money and go through all the stress and rigmarole. But I want to be his wife more than anything. We belong together, and I want to make this commitment to him and take his surname. It's just one day, I tell him; he'll enjoy it and we both know how much it means to our families.

Alison says the flowers have come. We gather around a white van in the farmyard to help carry buckets of sweet peas, stocks, scabious, delphiniums and roses through the garden and into the marquee. The clouds are beginning to break, and the last showers race away to the east. 'Wow, they look gorgeous!' I thank the driver and hand her a cheque. I run my fingers through the extra pots of lavender I ordered at the last minute and place both hands over my nose, inhaling the calming scent.

Back inside the tent it is damp and warm. I unfold the legs of a long trestle table, unpack my grandma's collection of old glass vases from a box on the floor and set them out. I ask Stuart to get some fresh buckets of water and my soon-to-be sister-in-law, Jane, starts putting the flowers into vases. 'There are ten vases for ten tables, three of these to go in each, and two of these and a few sprigs of this,' I tell her, waving different flowers at her. She looks nervous.

My old school friend Helen appears. I haven't seen her for months as she has been travelling. She hugs me excitedly, takes off her raincoat and says she is free all afternoon. She sets the chairs around the plain wooden round tables and, without prompting, finds a mop and bucket and starts cleaning the muddy floor. Anna, a friend I used to work with in the auction cafe, calls in to the house later on with a chocolate cake and a bundle of umbrellas – 'Just in case,' she says, smiling – then she's off to get her nails done, ready for tomorrow.

I haven't eaten or drunk anything in hours. Mum makes me a cup of tea and cuts the chocolate cake into deep wedges. It is gooey, dark and delicious. I take a moment to sit down in the farmhouse kitchen; amidst the chaos, I can see kindness and love all around me.

I think back to our announcement that we were getting married, at my dad's fiftieth birthday party. 'We don't want any fuss, a small wedding will be fine.' And when I suggested, 'What about going abroad?' – I had imagined a small wedding on a hillside in France or Italy – there was a sharp intake of breath. I saw instantly the worry and disappointment on my parents' faces. James's parents also looked puzzled. I knew they wouldn't want to travel, nor really be able to leave their farms. Mum and Dad declared that they would be paying for the wedding, that it was tradition, and that it should be a church wedding locally. I was caught. So, to try and please everyone, but wanting to hold on to my sense of self through the whole thing, I suggested that the wedding reception could be in a marquee in the field next to the farmhouse, so that I could do as much of it myself as possible. I'd make the invitations, and other bits and pieces, to look like those at the fancy weddings I'd waitressed at in Oxfordshire. I haven't bothered James with all the details.

Great-aunts and -uncles have dropped off wedding gifts at the farm all through the week before 'the Big Day', people I haven't seen in years. They console me about the weather as they sip cups of tea in the farmhouse sitting room and try to reassure Mum that even if it carries on raining it won't dampen spirits on the day. I open boxes of new pans and glassware, bundles of bedding and towels, vouchers for furniture, and money, gifts to wish us well in our married life. I sit on the side of the sofa near Mum and write a list of what everyone has given us. I will write thank-you cards afterwards, as if I am eight years old again, because the giving and the thanking are all part of the same ritual.

Back at the marquee I set up my old artist's easel with a simple seating plan. The whole place is starting to transform. Next to arrive are some freshly washed rubber cow-cubicle mats to lay over the muddy grass. Someone has found an iron and pressed the white tablecloths that hang over the tatty wooden tables. Little jars of Oxfordshire honey sit alongside little bars of locally made Penrith fudge. The summery flowers look amazing. I've hung up a board with a montage of photos of me and James as children on our farms, and our recent trip to Chile. Mum and Dad have made an area of comfy garden chairs with a coffee table in the corner for the old folk. Men from the local pub are setting up the bar.

I pin together the door of the marquee with the rope and wooden pegs, closing it until the big day tomorrow when I will walk in wearing my wedding dress, a married woman. I breathe a sigh of relief as I walk back to the house through Mum's pretty garden; she is brushing up the sandstone patio for the fifth time. As I look to the sky I can see the stars. The rain clouds have blown away.

My hair has been curled and pinned up at the back, leaving some strands down by my face. I pay the hairdresser as I leave, feeling all glamorous with my nails painted in a French manicure. As I drive back from town I can see the farmhouse standing on the hill, with the marquee in the field next to it glowing bright white. There isn't a cloud in the sky. Sunshine pours into the car. I start to cry: tears of relief that the sky is blue, and tears of joy.

I'm exhausted. I didn't sleep a wink last night and I spent the early hours sitting tucked up in bed, putting the final touches to an embroidery I have made of the farmhouse. I want to give it to Mum and Dad as a gift, to thank them for all their hard work and love in creating this day.

I wipe my tears so I can carry on driving, and I think of all the people that I will see today. I remind myself that if we had gone away to get married none of this would be happening. I am looking forward to one giant party in the sunshine.

I kick my shoes off and scoop the swishy princess dress around me to one side.

It was the first one I tried on several months ago when Mum and I went to a bridal shop in the nearby city one Saturday. It is 'swan-white' in colour, with little lines of embroidered diamantés down the bodice, and it makes my waist feel tiny. I have tied the fine satin spaghetti straps behind my neck in a bow. The full skirt has a similarly embroidered train that flowed behind me in a curve as I walked along the church path and down the aisle. It is now buttoned up with three invisible hooks and falls around me in a full skirt when I walk. I feel like Cinderella.

I sit for a moment on one of the wooden chairs to catch my breath. As I look around, my feet aching and my arms glowing red from catching the sun earlier, I think of James standing at the altar, staring straight ahead as he thought he wasn't supposed to turn around. I think of our laughter throughout the ceremony because we hadn't taken much notice of what to do in the rehearsal. I want to bottle this feeling, but it is nearly over. The lady I worked for at the auction cafe for seven years is sitting quietly at the table next to me. She leans over to say that I have made a lovely wedding and I should take a moment to look at everyone enjoying themselves. She tells me to take it all in, that it has been a wonderful day.

*

## Brussels Sprouts

Cooking Brussels sprouts in a steamer always takes me straight back to quietly sobbing behind the door in the little galley kitchen of our first house. A crowd of eager friends and family were sitting round tables waiting to tuck into their dinner. I blamed the hot steamy kitchen for my red blotchy face when I joined them. On that Christmas Day in 2004 when everything seemed perfect, it was all an act. I was empty, fragile and so very sad. It should have been a happy day filled with excitement, but for me it was a day to get through as best I could. We had just lost our first baby.

A few weeks before, we had found out I was pregnant. I was waiting for the twelve-week scan before we told anyone, because that was just what everyone did. I miscarried our baby ten weeks into the pregnancy. We were both shaken with the loss. We didn't tell anyone.

James and I were each other's safe place. And this was our first real test of looking after each other. It was hard for him in a different way; it was happening to my body, and my hormones and emotions were all over the place, but I knew he felt the loss acutely. I couldn't tell Mum what was happening to me as I started bleeding. That sad week she'd been called to Scotland to her estranged mother's hospital bedside. She had enough to deal with. It happened just days after I had invited the whole family for Christmas dinner. We'd planned to announce our happy news, and I had extended the invitation to a lot of family friends who we knew would be on their own unless they came to us. My brother-in-law had just lost his mum, aged forty-eight, to cancer, so we invited his dad and brother along too. I had to carry on. Our baby would have been the first grandchild on both sides of the family and first great-grandchild for my grandma.

For us, like for so many couples, my pregnancy didn't have a happy ending. But in my experience people didn't talk about miscarriage

– I didn't know of anyone who had lost a baby. I immediately thought there must be something wrong with me and that I'd never be able to have children. Miscarriage was a thing only talked about in hushed voices or behind closed doors; it was too messy, too inconvenient, too personal for everyday conversation. A doctor on the end of the phone told me to take paracetamol and get some rest, it would be over soon. Now I properly understood why women waited until the twelve-week scan before announcing their news.

I could have used another woman's shoulder to cry on. I should have reached out to a friend and shared my sadness and told her my fears. But, instead, I kept myself to myself. I walked and I walked, with our spaniel Bramble as my sidekick. She enjoyed racing around miles and miles of river paths as I tried to come to terms with not being pregnant any more.

James and I raced to finish the house renovations, the first of many goals we'd constructed, staying up late hacking bathroom tiles off, retiling, and painting ceilings. We laid flooring and I sewed cushions and hung curtains until the whole place looked lovely. On Christmas Eve I half-dipped shiny red apples into egg wash and then caster sugar for a snowy effect I had seen on TV, and I tied paper leaves to their stalks with names written in gold for each place setting. The table looked gorgeous as I turned the lights off and went up to bed. The house, the beautiful home, was all just a distraction.

From the moment of losing this baby, I knew I wanted to be a mother more than anything else in my life. It was like a gravitational pull. I had no idea what the reality of giving birth and becoming a mum would be, but it was something I desperately wanted.

I vowed to myself when I left Oxford that I wouldn't ever go back to working for someone else. I laughed it off with friends and said my CV made me look unemployable because I had only stayed in jobs for a short time. Now I wanted to become a mum so much that I didn't care a jot for any career plan. Jobs seemed irrelevant, a dull necessity to pay the bills. I had never experienced a desire so strong before. I became obsessed with my cycle and ovulation and anxiously waited, hoping to see another positive result on a little plastic stick in the bathroom. I saw people with babies in prams everywhere, and every advert on TV was about children or family mealtimes. I wanted a baby. I drove a nice car, I looked good and the house looked great. I wanted a baby. From the outside it looked like we were living our best life. But I wanted a baby.

## Spring Greens

The pink line is faint, but it is appearing as I wave the stick and hold it up to the light of the bathroom window. I hold it in hope and disbelief, three months after miscarrying. I am pregnant again.

I take a deep breath and realise my hand is shaking. I sit down on the loo seat and wish I'd bought another test to double-check. A mixture of anxious excitement floods through me; I am elated and terrified at the same time. I don't want to put James through an emotional rollercoaster again, but I have to tell him. He won't be home for hours. I look at Bramble, who has made her way into the bathroom to find out where I am. She wags her tail at me and I bury my head into her, tears welling up in my eyes. I wipe them away to look again at the plastic stick, just in case the pink line has disappeared. It is still there, slightly deeper in colour than a moment ago. Life is so fragile. I am holding proof of my pregnancy in my hands, and I have a little cluster of cells inside me, attaching themselves right now to the lining of my womb, growing into a baby.

All the 'what if' thoughts race through my head. I am still grieving our loss and this result seems like a miracle. This pink line is a new life. I know it could all go wrong again any moment, and for so many women it sadly does. But equally it could mean that in nine months' time I will become a mum. I try to slow my mind down and breathe, remind myself to be calm. I am only six weeks along and another six weeks until I can have a scan seems like an eternity. Even the next three hours feels like forever. Our loss is still so raw.

I text James, tell him not to be late tonight and then go for a long walk with Bramble. It is February and bitterly cold, but a walk is what I need. I am pregnant. I am pregnant. I am PREGNANT.

When I get back home I turn the heating on and cook us a big bowl of noodles with beef and fresh greens. James is barely through the door when I tell him to look at the stick. He hugs me at the sight of the little pink line and my news, but I can see he is he guarding himself, not wanting to trust in this bit of plastic. He thinks I should do another test in a few more days. He doesn't want to invest all his hopes again so quickly. He fiercely wants to protect me and warns me against getting my hopes up too high, but I want to shout it from the rooftops: We are having a baby.

Any remaining interest I had in my work disappears overnight. I'd started a promising small business, finding houses for clients. I worked from home and all I needed was my car, my laptop and good relationships with the local estate agents. I'm good at it, because I know the area really well, and I love matching people with the right houses. But I'm just not interested in it. Instead, I become obsessed with eating well, getting plenty of fresh air and keeping the house in order. As my belly grows and the scans are all OK, I focus on being fit and strong ahead of giving birth. It is my sole purpose now. I still have a few contracts to fulfil with clients and I do the work, but it is just a means to an end.

SEE P.302 FOR **NOURISHING FOOD FOR SELF-CARE**

# BEEF NOODLES WITH GREENS

Prep 15 minutes

Cook 15 minutes

Serves 4

## Ingredients

150g, or 3 blocks, egg noodles

1 small head of broccoli cut into florets, or 300g/10oz tenderstem
broccoli

1 tbsp sesame oil

1 x 400g/14oz beefsteak, sliced – rump or a thin minute steak or any
other cut you can get (or 400g/14oz beef stir-fry strips)

2 handfuls fresh beansprouts or 225g/8oz tinned water chestnuts

1 head of pak choi, sliced

2 spring onions, sliced finely

## For the sauce:

2 cloves of garlic, crushed

2cm piece of fresh ginger, grated

3 tbsp soy sauce

2 tbsp oyster sauce

1 tbsp white wine vinegar

1 tbsp tomato ketchup

Note: Mushrooms also go well in this dish instead of beansprouts or
water chestnuts; at the end of step 3, remove the fried beef from the
pan, fry the mushrooms, then combine the two and add the sauce
and pak choi.

## Method

1. Cook the noodles in a pan of boiling water and simmer for 5 minutes
   (or according to pack instructions). Add the broccoli 2 minutes before
   the end of cooking the noodles and drain. Rinse under cold water
   and set aside.

2. Combine the sauce ingredients in a small bowl.
3. In a wok or large frying pan, heat the sesame oil and stir-fry the beef on a high heat for 3 minutes until browned but not cooked through.
4. Add the sauce to the pan with the beansprouts/water chestnuts and pak choi for 1 minute before tossing in the noodles and broccoli.
5. Scatter with fresh spring onions.

James and I have regular conversations about money, and they always end in heated rows. He asks why I'm not advertising more and pushing for new clients – the business has so much potential, and the people I've looked after were really happy with the service I provided and recommended me to their friends. I could grow it and make some serious money. He's frustrated with me because he knows I want none of it. In a few short months I'll be nursing a baby, and I don't think he's thought things through properly. Who's going to take care of the baby if I'm off working? I'm certainly not going through all of this just to hand him or her over to someone else to look after. And why would I want to be stuck at a desk or making calls instead of taking care of our baby?

But I still keep a close eye on the property market, and start to see houses on our street selling for nearly double what we paid for ours. We've done a lot of renovation work, had the place rewired and installed a new kitchen and bathroom, so we decide to put it on the market and look for somewhere with a garden.

We're out walking one weekend and see a 'For Sale' sign going up in a village on the edge of the city. I cross the road when the van drives off and peer through the windows of this empty, dilapidated old millworker's cottage. It's in a terrace and all the other houses around it have been done up. I phone the estate agent when we get home and organise a viewing. I know a place like this will get snapped up quickly. We wander around the empty rooms; it's like a museum, with an old scullery and coalhouse and little old fireplaces in the three bedrooms and the front room downstairs.

The garden is the best bit for James, and Bramble – completely overgrown and neglected at the moment, but perhaps fifty feet long. There's an old apple tree at the end of it.

The legalities go through without any problems, we soon have the keys to the cottage and are able to start work on it within a couple of months. Enough time before our baby arrives. Managing the build and all the permissions and regulations and the sale of our own house, packing up to move and decorating the new place keep me busy. I have no time to be looking for any more clients, and besides, the money we made on the sale of our first house is more than enough for what we need to renovate the new one. For a while, the rows subside.

I am standing on a stepladder and I know I shouldn't be. I am five months pregnant and scraping wallpaper off in the front bedroom. Layers and layers of old paper, starting with gaudy blue stripes, then green fleur-de-lis, a faded floral ditsy print, a pale pattern of leaves that I like the most, then bare plaster. Some of the pieces come off satisfyingly, like peeling big patches of burnt skin off my arms when I was a girl. Some bits are stuck fast, and I have to scrape harder. I see the history of the house in these layers. I imagine the people who lived in these rooms. The most recent paper is the worst, bold blue lines, enough to turn anyone into an insomniac. Bramble lies at the foot of my ladder – it's only three steps high but she likes to stay close by. Helen calls in to help and forbids me to climb any more ladders. Between us we get the walls stripped, lined and painted. It will be the bedroom I bring our baby home to when we move in.

The shape of my days is well set. I do my office work in the mornings at our other house, and then walk here with Bramble along the old millworkers' path by the river. I carry a bag with me with fresh milk for a cup of tea and some sandwiches or a salad for my lunch, and work in the house all afternoon. James comes to pick me up after work to see the progress. It's always empty and cold when I get there, but I warm up as I start working.

The green front door has a little brass keyhole and no handle, just a metal catch, so I am careful not to lock myself out. As I walk inside there is a flight of stairs straight ahead, and a square front room on my left with an old tiled fireplace and a door into the little kitchen that looks like the room time forgot, a washboard standing on the floor next to the stained Belfast sink set on bricks with old copper taps. The bathroom is just through the kitchen in a strange arrangement; it has a cast-iron bath in green and a tatty old shower curtain badly attached to the ceiling – one pull and it's down. I look at the concrete floor and think of the old lady who lived here for a long time. I imagine what it was like to scrub clothes on a washboard and cook on the range that would have been in the fire-place. Her house came up for sale after she died and I think about how she lived here like this with no central heating or modern comforts, just bare stone on the kitchen floor. Until recently she had to carry coal from the coalhouse, still half-full at the back of the house. And I think how odd that her bedroom was papered in blue-and-white stripes. I love that houses can tell stories. I want to know who lived here before her, in the terrace that was named after the mill owner's daughter, Margery. This house would have been home to workers who wore clogs and bathed in a tin bath in front of the open fire. The builders soon have it all stripped out and they fill the yellow skip out front. They work fast to create a large extension with patio doors that open onto the long garden.

We drive to the farm together one Saturday morning in mid-November. Bramble leaps out the car boot and I attach her lead to her collar as she circles around me. I pull my cosy hat on, the wind whipping around us, and wrap my scarf tightly around my neck and under my coat. I walk with her slowly, more like a cautious waddle than a walk. I take her along to the little humpback bridge past the Oak Tree Field; she doesn't pull me. I let her have a run around in the corner field on the way back. James is checking the flock with his dad. I am due in three days. They both tease me, saying I need a bumpy ride on the back of the quad bike to get this baby out.

On the way home we stop at a farm shop. It has a gallery attached and we decide to buy a simple painting of a golden field of barley. We haven't much money, but we love it, and it feels like something we will keep forever. I unpack it back at the house, wandering around thinking about where to hang it. James puts the TV on. He has been listening to the rugby match on the radio all the way home. I spend the afternoon ironing, folding and putting away clothes. I iron shirts that have been sitting in the basket for ages. I can't sit down. I am pacing around the house finding anything to do to make it look cleaner and more organised. I go for a bath but struggle to climb out and call to James to give me a hand. I feel like an elephant trying to scramble up a riverbank. And then it starts: the waves of tightening over my abdomen. I think I'm in labour.

'Don't panic!' James says.

'I'm not panicking!' I snap back, but I need to finish ironing those sheets and I want to change our bed.

He tries to tell me to stop working, but I am not listening to him. The midwife has told us that we shouldn't go to the hospital too soon, and I get the impression that he doesn't mind if I hold back a little because England are close to beating New Zealand and he can't take his eyes off it. I eat a bowl of cereal and tell him he can help himself to cheese and crackers or something for his supper. I haven't got an appetite for a meal. I go upstairs and rest my legs by sitting on my birthing ball, unable to lie down or stand still; I rock as each surge comes through my body. I check and repack my hospital bag and decide to call the hospital and tell them we're coming in. 'We need to go now.'

James is quite excited as the match is in its final minutes, but I am not interested. As I get in the car in the driveway, he is hovering by the door, watching the screen to check the score. I shout,

'Hurry up, the baby is coming now!'

He seems alarmed. 'Christ, why didn't you say so?'

I am standing at the side of a bed in a dimly lit birthing room. I can see a crib made up. My cardigan is draped around my shoulders and I breathe deeply. My contractions are close together and take over my whole body. I bury my head into the bed, which has been raised. My legs are planted firmly on the floor. James is by my side. He holds me, rubbing my back, and I grip his hand. I give everything I have and more, lost in my breathing, punctuated by calls of 'Come on, keep pushing.' An hour after arriving in hospital we have our baby girl. She is 7lb 5oz and has a shock of dark hair. I am helped into the bed to rest, elated, hungry and streaming with tears. The midwives pass baby Molly to James, who looks startled and completely unsure of how to hold this wriggly little person, but he is grinning from ear to ear. He holds her like she is the most precious thing in the world, which of course she is. She gazes into his eyes and he sings 'Baa, Baa, Black Sheep' to her.

## Cold Tea

I bump the wheels of my new Silver Cross pram down the cobbled path to the river every morning. We follow a track that cotton-mill workers would have walked every day to the factory. Bramble is by the side of the pram and doesn't run off like she used to – it is as if she is protecting her own baby. I always walk back home by the road and as I reach the top of the hill I am out of breath. I lift the pram up our three steps and park Molly up by the front door, leaving it open as I nip into the kitchen to make a cup of tea. There is a little primary school opposite our house and the children are usually out in the yard playing at this time. I want to sit for ten minutes with my hot drink and rest, but the crying always starts five minutes after we get back. I unwrap her like a parcel out of her cosy snowsuit, trying to comfort her, and go inside to start the feed–change–play routine again.

James is racing between three jobs to pay our bills. He has picked up regular consultancy work, is doing some freelance projects, and the farm pulls him back. He is out the door before seven each morning and comes home hungry and tired every evening. After Christmas I am approached by a couple looking for a house and I take their search on in a token effort to keep my business going. I take tiny Molly with me, strapping her to my front while I view properties and take photographs. Bramble rides along in the boot of the car and is always happy to have a walk by a lake or woodland path on the way. It all seems flexible enough to everyone around me, but the juggle is starting to stress me. Even just lifting the buggy in and out of the house and into the car is too much. I am irritable about little things and feel frustrated all the time. Playing two roles – 'a professional woman' and 'a mum' – isn't something I can or want to do. I want to be a mum, and I want to do it 100 per cent.

Molly is an uneasy baby. I barely get any proper sleep for months. James offers to take her out in the car or for a walk at 4 a.m. so I can try and sleep before he goes off to work. I suggest he sleeps in the spare bed, but he doesn't want to leave me, and we end up with Molly in between us every single night, and then I don't sleep for fear of him rolling onto her.

Because I wanted all of this so much, I hide from everyone the fact that I'm not coping. Every day I put on clean clothes, wash my hair and try to remember to put on a little make-up; I smile, cook home-made soup and bake cakes for visitors. Our house is immaculate, and I am falling to pieces.

SEE P.302 FOR **MEALS FOR WHEN I'M IN SURVIVAL MODE**

Six weeks after Molly is born, she isn't feeding well. My nipples are badly cracked and bleed when she tries to latch on. She is losing weight. When the health visitor comes to do her check-up, she helps me apply special medical gauze used to treat burns to my cracked skin. It starts to soothe me straight away. She sits down with me after weighing Molly and writing a lot of notes in her book. She asks about my routine as I tuck a nappy back on my baby's little pink body and dress her carefully. I explain that she is always unsettled after her feeds and in order for my husband to get some sleep, because he has to get up for work, I often take her downstairs and sort the washing or watch TV while I rock her through the night. I walk around the house carrying her, desperately trying to settle her back to sleep, softly singing 'A Hundred Green Bottles Hanging on the Wall'. It doesn't work – she just cries at me all the time.

The health visitor tells me I should keep the night feeds as quiet and as much in the dark as possible, feed her when she is barely awake and try not to disturb her too much, only changing her if she has a dirty nappy rather than just a wet one. I tell her that my mum told me that when I was a baby I just fed every four hours and slept like a dream in between. I am trying to hold it together but as she places her hand on my arm I feel my shoulders drop. 'What am I doing wrong?' I ask her with tears rolling down my face. She smiles, handing me a tissue, and tells me that babies are all different and that I shouldn't be so hard on myself, and that my mum has likely forgotten about the tougher times.

She suggests that I try to fill Molly up with a bottle. She asks me if I'm eating enough.

'Not really,' I say. 'I often skip breakfast to walk the dog and when I get back I'm starving so I eat chocolate bars, cake or biscuits, then I don't have a proper lunch. I try to make a proper meal in the evening because James is home, but I usually can't face cooking and we end up eating cereal or junk instead. Our fridge is full of Lucozade and ready-made pizzas.'

'You have to eat properly,' she says firmly. 'Don't miss breakfast. You need foods like porridge, yogurt, sausage, bacon or eggs in the mornings. For lunch you need to fill up on good brown-bread sandwiches, or soup, rice, chicken, fish and some fruit. You need protein with every meal to keep your energy levels up.'

'I know, I'm really cross with myself. I've let this slip. I ate like that while I was pregnant. My walks are just about holding me together but I know I need to meet up with other mums and I need to see some friends. There's another mum that I met at an antenatal class – I'll call her.'

'You need to stop stressing about keeping the house in such perfect order,' she instructs. 'Leave the bloody cushions alone and grab some rest through the day whenever you can.'

We dream of moving to the farm but there are so many complications and hurdles to overcome. With hindsight I can see that from the start we put far too much pressure on ourselves. The present was never enough – we were always chasing the next big thing. James is constantly stressed and unsettled. I can't see a way of making it happen and I desperately want to enjoy our life *now*.

From the outside it all looks perfect. We have a gorgeous home. The tatty front door has been painted. We have planted a pretty front garden, built a brand-new extension at the back, laid a lovely lawn, fitted a new kitchen and bathroom, and we have a stylish limestone fireplace in the living room. It is all light and clean. I've put matching cushions on twin sofas and we have neat bedspreads as if we live in a show home. I have made it into a lovely space to be in. But I am miserable. 'Having a baby' in my mind and 'having a baby' in reality are two very different things. The surge of emotions, my fierce desire to protect her, feed her and comfort her, and keep our house immaculate, and look like I have it all together, is draining me. Before the health visitor leaves, she asks me to fill out a questionnaire about my thoughts and my mood. I don't know this at the time, but looking over my answers, she tells me later, she is deeply concerned about my mental health.

# 4

## Chocolate Buttons

Molly has started moaning, her little arms waving for me to pick her up. She is in her baby walker, stranded in the middle of the dusty kitchen floor, while I make five cups of tea for the workmen. Here we are again, a few short months after the last house was finished, living out of boxes. The kitchen walls are stripped to the bare stones. The plaster was hacked off last week, and the electrician wants to be in next to do the new wiring. She is ten months old and not quite walking. I spend most of my days straddling over her, holding both her hands as she learns to put one foot in front of the other. She hates the baby walker and just wants me. I give her a piece of banana to pacify her, but she throws it onto the dirty concrete floor and howls.

I have a battered old kettle set in the kitchen windowsill, plugged into an extension cable from the only socket that still works in the next room. The water is from an outside tap. I seem to be constantly driving back and forth from my mum and dad's to the new house with Molly, timing the journeys to fit in with her naps. When I get here I make tea, sweep up or get on the phone to order more supplies for the builders. I call round the builders' merchants for the best prices on blocks, insulation and wood. I compare prices of guttering, flooring and tiles and order paint. We are doing another 'project' and most of the time I am fizzing with the excitement of it all – I love being the project manager and turning a place from a dump into a beautiful home.

James is working in a job he hates, but we need the money. He leaves every morning before Molly is awake to shepherd at the farm before changing in his car and going to a desk job. He is usually back to the farm after work and doesn't get home until she is nearly in bed. Most families must live like this, I think, when I feel weary from doing all the childcare. We have just stretched ourselves as far as we dared to buy a house to be nearer to where we both grew up and to be closer to the farm. It feels like every previous generation just somehow ended up living where they always had. Nowadays that seems impossible. We are pushing ourselves to the limits to make our lives happen. On good days we are full of confidence about making our own success. I tell myself regularly: Who wants to move into a place that's all done up anyway?

This is our third move in three years. The house is in the middle of a little village. It is a traditional Cumbrian longhouse. It was built for a farm manager to live in on a little estate that once owned the land round it. The original inhabitants would have worked for the squire, who lived next door in the grand manor house, on the other side of the big wall that we look onto. Our kitchen is in the end of the house that would originally have been where the animals were housed. Cattle and sheep would have been bedded up in here through winter, providing warmth for the grain store above, which is now the bathroom. The house is a major step up the property ladder for us, and we could only afford it because it was so trashed, and because the survey described all sorts of problems.

I peer out of the dirty window over two squares of mossy lawn that sit either side of a path from the front door, which has the date 1692 inscribed in stone above it. Rusty black-and-white railings enclose a scruffy front garden; they need scrubbing with a wire brush and repainting. The little gate opens, with a creak, onto the road that passes through the village. There is a stone-built bus shelter across the road and as far as I can tell the public bus only comes to this village about once a month. Swallows sweep in and out from their carefully built nests in the bus shelter and away across the village green. The green is mowed by an old man called

Sam who doesn't seem able to stand upright. It is bright and sunny outside.

A radio is blaring nineties songs in the room next to me, where two builders are loading up and taking wheelbarrows of plaster out to the yellow skip parked in front of the house. They burst into song when they like a tune. In one of the bedrooms above me another old man is working on the ancient sash windows. He likes to shout instructions at the other guys, as if he is the foreman, and thinks he needs to keep me right about the work we're doing, assuming I'm too young to know anything about old houses. 'This is three-millimetre picture glass,' he tells me as he kneads putty in a dish with an old kitchen knife. 'Do you know they're called Yorkshire lights, these windows? They couldn't get glass in any bigger-sized panes back then.' He has been working on historic houses like this since he was a boy, and he has all the tools and know-how to fix old things. He fills his days with odd jobs like this, and tells everyone who will listen tales of his working life, like the time he was told to get down off a ladder on a street as the Prince of Wales's convoy was passing through. He refused and said he didn't have time to stop working for that nonsense, and then the prince slowed the car down to stop and have a chat with him about what he was doing.

There are two yew trees badly trimmed into teardrop shapes either side of the largely unused front door, which has cobwebs surrounding the frame on the inside. I imagine next summer, looking out of my kitchen window onto a trimmed lawn where I can set a picnic blanket for Molly to play on, the path now flanked by a lavender hedge that I will fragrantly brush against as I walk down it, and it will be surrounded by beds full of delphiniums, roses and geraniums. I plan to paint the front door a soft sage-green colour. This house and its little garden may be a dump at the moment, but I know I can make it the house of my dreams.

The walls inside will be painted in light duck-egg blue or cream, and the windows will be dressed with curtains I have made. I want to put coir matting on the floor downstairs, like all the posh country houses have, and soft pale wool carpets upstairs. We can have a Victorian-style roll-top bath, with tongue-and-groove panelling around the walls of the bathroom, just like I dreamed of back in Oxford. The deep windowsills will display vases of freshly cut flowers or family photographs or the large washbowls and jugs I've collected at vintage sales. I will cook our favourite dinners in the kitchen and James will be playing with Molly in front of the fire, Bramble snug in her dog bed next to them. And out the back window of the kitchen I imagine a tidy backyard stacked with logs.

I am waiting to meet a kitchen planner from our local joinery company to help me decide the layout of my new space. He's late. I can't keep Molly entertained for much longer before she'll be screaming for her lunch and a nap. The sound of the cement mixer chugs away in the backyard. This doesn't yet feel like my house – these men are all over it, working away, knowing more than I do right now about how to fix it up – but one day it will be mine. The builders are rebuilding part of the back wall of the house that has subsided over the years. I spy a car pulling up next to the skip so I carry Molly through the house and pull the stiff front door open to meet the guy who has come to measure up. I juggle the conversation with my restless toddler on my hip and smile throughout as if I am managing fine.

I show him inside, past the pile of rubble and dust, being careful to step around the dehumidifier in the room. He laughs: 'You've got your work cut out here, luv.' I look around and my usual optimism wavers; in that moment I feel completely overwhelmed by the amount of work and expense ahead. The carpets are flea-ridden, and upstairs the bedroom walls are all badly painted in deep purple and dark green. Old chip fat is stuck fast to the kitchen ceiling so we need to pull it down and start again. I try and hold on to the excitement I had when we viewed the house several weeks ago. Then, I looked past the stained bathroom suite, mouldy shower room and damp patches on the ceilings. Now, these problems are all that I can see. This is a lousy place to have a ten-month-old baby – everywhere is unsafe. She wants to crawl, pick things up, put things in her mouth. It won't be fit to live in for months. I told my mum and dad we'd only need to stay a fortnight, and that was weeks ago already.

Talk of the new kitchen lifts my mood. I discuss the plan with the man from the joinery and choose oak worktops and cream-painted units from his samples and a catalogue, in a basic one-length layout to keep the cost down. Molly is nervous around strangers and clings to me like a young orangutan in one of those David Attenborough documentaries, watching him with her beady eyes. She tugs my hair and tries to cover my mouth with her hand while I talk. I keep the conversation brief and to the point. He doesn't get far with wanting to sell me additional wall units or cupboards that I don't need, and I don't offer him a drink. I see him out. I need him to go so I can sort out our lunch.

SEE P.303 FOR **EASY WAYS TO FEED LITTLE ONES**

Feeling flustered, I have to put Molly back in her baby walker to sort out the food I have. I know she will start to scream as soon as I put her down, but I have no choice. After being offered a bread-stick or two from my bag, she calms down. I warm up a tub of last night's leftover pasta in the microwave that I brought here for the builders to heat soup in – little macaroni shapes with passata and bacon stirred through it. I give her some pieces of cheese and sticks of cucumber while it cools slightly. I make myself another cup of tea and have some ham sandwiches out of my Tupperware box. Molly tries to feed herself but is getting covered in the red sauce, so I take over and spoon her little mouthfuls. Bramble is drooling, waiting for any bits we drop.

After a quick clean-up with some baby wipes, I carry Molly to the car and change her nappy on the back seat; she twists around and tries to escape. As I unload her buggy from the boot, Bramble is wagging her tail around me. She has been very patient until now. We go for a walk up the road to the duck pond and then around a track by the neighbour's farm. Molly watches the black-and-white dairy cows pulling grass up into their mouths. They are curious and lean over the gate and breathe through their hot wet noses at us. The sun warms my face and I smile at Molly and pass her a packet of chocolate buttons from my pocket.

A few weeks later, feeling smug with my low-cost kitchen all booked and paid for, I completely lose my mind when I see fancy limestone flags in a showroom and splash out on ordering them for the floor. I phone James to tell him that we can save money by laying, grouting and sealing them ourselves. He complains about them a few weeks later because it turns out they take many nights to cut, lay, grout, seal and seal again with the special protective liquid. They look amazing and lighten the whole space up, but over the next seven years I discover they are a nightmare to keep clean. (I'm careful never to mention this to James as he'd have definitely bought the cheaper, more practical ones.) My other big extravagance for the house is the cooker, a cast-iron Rayburn range in classic cream enamel that we will fit in the old sandstone fireplace in the kitchen. There is no way we can afford this, but to me it is the heart of the home – I can't do without it, I tell myself. The builder shapes sandstone into curved shelves either side of it and it looks wonderful. But our oil bills will be ridiculously expensive. Over the next few weeks I wash, scrub, paint and polish, working hard to create my own farmhouse-style kitchen. I set my book-case/dresser in the corner and hang an old set of pine shelves to show off my favourite pottery. Our wooden table and chairs from Oxford sit neatly under the window. We call in decorators to paper and paint the big square bedrooms, borrowing extra money to finish the house. The last two houses we did up like this more than paid for all the refurbishments when we sold them. That's the logic

I'm using now to justify our mounting debts, but I don't want to sell this one.

We have been storing all our furniture in my grandma's garage. She likes it when we call in to pick something up or drop things off. Every time she wants us to stay for longer than we have time to, to sit and chat. She lives on her own and clings to any visitors like a limpet. She pours James a whisky and tries to fill my glass with a sherry from the silver drinks tray she keeps on top of her polished-mahogany cupboard in the sitting room. I tell her I can't drink as I'm driving. 'Oh, a little one won't hurt you,' she says as she fills my glass.

Today, Grandma sits with her warm hands crossed in her lap, smiling as she watches Molly play. I have set her up on the floor with a bag of toys I fetched. Molly pushes little cars and stacks coloured rings onto a wooden stick. After a couple of visits like this, I suggest we leave a little wooden box on the hearth with a few things just for playing with at Grandma's house, so I don't need to remember to fetch some each time. Grandma asks me the same questions she did the last time we came, about the house and when it will be ready. She tells me about going to furniture sales and buying things for the farmhouse when they first moved in. 'The china cabinet behind you came from my mother,' she says, 'and that tea set was a wedding present to her – would you like it when I'm gone?' She almost insists I take it there and then, but I say, 'That would be lovely, but it looks so nice in there, let's leave it for now.'

I set Molly on the floor with the two doors open so I can hear her from the kitchen, and start to make Grandma a cup of coffee with a tiny drop of cream in it, the way she likes it. I wash up the lunch things on the sink while the kettle boils. She seems to eat the same thing every day, a bit of cold ham or cheese and bread and chutney, a pear and a coconut tart or bit of fruitcake. There is a lamb chop defrosting on the worktop next to a pan of cold water with a couple of peeled chopped potatoes in it, and a bit of cabbage laid out ready to cook later. 'Do you want me to do anything for you while I'm here?' I call through. She says I can get the sheet in from the washing line because her knee is getting worse and she finds the steps tricky. I know she wants to talk some more when I come back in, but I don't really have the time. I am rushing to visit her between Molly's feeds and naps and getting back to the house to meet the plumber.

Grandma's days are long and empty. The minutes slowly tick by on the clock in the hallway.

We are still living at my parents' and I am desperate to get our house finished and move in. They are equally desperate (and don't need to say it) to have us move out. We set ourselves a tight deadline of moving before Christmas. The carpet fitters are the last to come in, and then we can set our beds up. I spend my evenings making curtains when Molly is asleep. I take her to choose a few new cushions to dress our old sofas up, even though we are out of money. She takes her first steps in a large home-furnishings store and I clap my hands as she toddles past the staged room set-ups and fetches me rolls of wallpaper samples she has pulled out of a box. I want to make a home for the three of us to somehow help James feel better about the hours he is spending away from us doing a job he despises to pay for it all. Every evening, weekend, and any holiday James can take from his job, we sand skirting boards, paint, lay tiles and pull nails from the old ceiling joists. We ask endless favours of friends and family, to help with the jobs or look after Molly. Making our way in the world, in reality, involves a lot of other people.

Through the months of renovations, I try and cook meals at Mum and Dad's to help out – they like my lasagne. But I never know when James is coming back so we eat at different times. Mum is doing her best to accommodate us, but I don't want to impose more than we have already. Her fridge is taken over with Molly's little tubs of pureed vegetables, yogurts and bottles of baby milk. There are two kinds of butter: Mum's favourite low-calorie spread and ours, real butter. And I buy curry pastes and hummus that they don't eat so it leaves no room for Dad's cheese. These little things start to add up. Mum's work surface is cluttered with the steriliser, blender and my food processor. Our washing is never-ending. Mum puts it through with theirs, hanging all the wet things on racks, and then irons everything because she doesn't have a dryer. I feel guilty about all the extra work. There are plastic toys in the bath and coloured blocks and crayons on the floor. Dad mutters under his breath when he can barely find a coat peg to hang his jacket on. The buggy is always in the way and there is always someone in the bathroom. The plan to move closer to family has put a strain on us all. James and I like to think of ourselves as self-sufficient as a couple, but now we have my parents' first grandchild, and we live under their roof, our lives can be commented on constantly. All the decisions we have to make are debated. It's noisy and there is regularly a bad atmosphere. I can't wait to leave, because trying to keep the peace between us all is so incredibly stressful. I know the last thing James wants to hear after a long day is me moaning about how Dad chucked out the leftover rice pudding that I had saved for him.

## LASAGNE

### FANCY VERSION

If you can, make a large quantity (i.e. double the recipe below) of bolognese sauce and freeze some for another day. This can be made 4 days ahead of when you make the lasagne, stored sealed in the fridge. My béchamel is made by adding all the ingredients in at the same time and whisking until thick – I don't add cheese to my sauce as I prefer to scatter some on top of the lasagne before cooking.

Prep 45 minutes
Cook 1 hour 30 minutes

Serves 6–8

Ingredients

For the lasagne:

2 tbsp olive oil

250g/9oz beef mince

250g/9oz pork mince

salt and pepper

2 garlic cloves, crushed

1 onion, finely chopped

3 sticks of celery, finely chopped

1 carrot, finely chopped

50g/2oz pancetta or streaky bacon, diced and fried (optional)

5 button mushrooms, sliced (optional)

1 tbsp tomato puree

1 x 690g bottle tomato passata

1 glass red wine

200ml/7fl oz beef stock

50ml/2fl oz whole milk or a swirl of single cream

500g/1lb dried lasagne sheets (you may not need all of them depending on the dish you use, and some may need to be broken to fit)

For the béchamel sauce:

50g/2oz butter

50g/2oz plain flour

570ml/1 pint whole milk

Parmesan or Cheddar, grated for sprinkling

Method

1. In a large heavy-bottomed frying pan, add 1 tbsp of the oil and cook
   the mince until browned. Season well while it is cooking.

2. Transfer the mince to a slow cooker or casserole pot.

3. Add the remaining olive oil to the frying pan and then add the
   pancetta and vegetables, apart from the mushrooms (if using). Sauté
   them for 15 minutes over a medium heat, until soft, adding a little
   water if they start sticking.

4. Transfer the vegetables and pancetta to the slow cooker or casserole pot with the mince.
5. Add all the other ingredients (tomato puree, passata, red wine, beef stock) except the milk/cream. Stir well.
6. Leave to cook for 3 hours on a low heat adding the mushrooms in the last hour.
7. Stir the milk or cream into the meat mixture 10 minutes before the end of cooking. Remove from the heat and leave to cool.
8. Make the béchamel. Add all the ingredients to a pan and whisk over a medium heat for 5–10 minutes until thickened. You want a silky sauce rather than gloopy. Season well.
9. Heat the oven to 200°C/fan 180°C/gas 6. In a large baking dish or tray – roughly 30 x 30cm – layer up the cooled bolognese sauce with lasagne sheets and top with the béchamel sauce and grated cheese. You'll get three layers with this quantity of sauce. Bake for 40 minutes until golden and bubbling.

Note: Chefs making béchamel would heat the butter in a pan, add the flour and whisk to form a 'roux' and then add the cold milk little by little, whisking over the heat until the milk is combined into the flour and butter and the sauce has thickened.

CHEAT'S LASAGNE

You can make a faster bolognese sauce for your lasagne by browning the mince and adding a ready-made jar or two of pasta sauce and simmering for 20 minutes. I would recommend adding 50ml of milk or cream towards the end to enrich it. Make the béchamel sauce as before, and layer up and bake the same as above.

# RICE PUDDING, THREE WAYS

Serve with a quick home-made raspberry jam (see below).

## SIMPLE

**Prep 10 minutes**

**Cook 2 hours**

Serves 6–8

### Ingredients

a knob of butter to grease the dish

100g/4oz short-grain pudding rice

100g/4oz caster sugar

1.2 litres/2 pints whole milk

### Method

1. Heat the oven to 170°C/fan 150°C/gas 3.
2. Grease a 20 x 20cm baking dish.
3. Rinse the rice in a sieve under cold water.
4. Put the rice, sugar and milk in the baking dish and stir together.
5. Bake in a low oven for 2 hours – stir a couple of times in the first hour, but then leave to form a brown skin.

## SLOW COOKER

This cooking method won't form a browned skin on the rice pudding but is ideal if you don't want the oven on for ages.

**Prep 5 minutes**

**Cook 4 hours**

Serves 6–8

### Ingredients – as above

### Method

1. Grease the bowl of the slow cooker and add the rice, sugar and milk.
2. Place the lid on and cook for 4 hours on a high setting (or until the rice has absorbed all the milk), stirring halfway through cooking.

# LUXURY (CLOTTED CREAM RICE PUDDING)

**Prep 15 minutes**

**Cook 1 hour 35 minutes**

Serves 6–8

## Ingredients

a knob of butter to grease the dish

850ml/1½ pints whole milk

280ml/½ pint double cream

227g/8oz tub of clotted cream

1 vanilla pod, split

100g/4oz pudding rice

100g/4oz golden caster sugar

fresh grated nutmeg for the top

## Method

1. Heat the oven to 180°C/fan 160°C/gas 4. Grease a 20 x 20 cm baking dish.
2. Put the milk, double cream and clotted cream in a large pan with the vanilla pod and bring to a boil – be careful not to let it boil over.
3. Remove from the heat and add the rice and sugar, stirring well. Take out the vanilla pod, rinse gently and dry to reuse.
4. Transfer the hot mixture carefully into the baking dish and even out the rice with a fork or a spoon.
5. Bake for 20 minutes, then reduce the oven to 160°C/fan 140°C/gas 2 and bake for a further hour until the top is brown and bubbling. Allow to cool slightly before serving with a grating of nutmeg.

# A QUICK (TEN MINUTE) RASPBERRY JAM

Makes 1 small jam jar

## Ingredients

100g/4oz caster sugar

2 tbsp water

225g/8oz fresh or frozen raspberries

**Method**

1. Heat the sugar and water in a pan until the sugar has dissolved.
2. Add the raspberries, breaking the fruit down with the back of a spoon or fork. Boil for 3–4 minutes until it has formed a jam – it will thicken as it cools. Can be stored in the fridge for 3–4 days.

Bramble is also a point of contention. Mum loves her, but she does not love her wet muddy spaniel paws. Two weeks has turned into three challenging months. When we eventually pack up and leave, she calls me a few days later to tell me that she has deep-cleaned her house from all that horrid dog hair and insists that I get Bramble clipped more regularly. I snap back, saying I haven't had a haircut in months, never mind the bloody dog.

I love making food for the three of us when we're working at the house. Porridge, bacon and eggs, pasta – basically, anything that can be cooked in a pan on the camping stove. We also eat a lot of bread, cheese and cold cuts. The fridge is hidden under a dust sheet to keep it from being splashed with paint. I make copious mugs of milky builder's tea from the paint-splattered kettle. When the kitchen is finally complete and the Rayburn switched on, I celebrate by making roast chicken with a rhubarb crumble for pudding and invite Mum and Dad over. Everyone is relieved. I bake trays of gingerbread and caramel shortbread, dropping them off to friends to repay their kindness. It is Molly's first birthday in November and all I want is some quiet days.

Just before Christmas we host a housewarming party, inviting everyone who has worked on the house to thank them. The house looks better than I ever imagined it. One day only a few weeks ago, when we were hacking plaster off, a huge lump fell and revealed the original fireplace in the sitting room – a carved stone fireplace from the seventeenth century that was similar in design to the carved date stone above the front door. It was a fleeting moment of joy, making us feel like all of this dirty renovation work was worthwhile.

For the party, James trims and decorates the two yew trees either side of the front door with fairy lights. The plumber is busy fitting the last basin and toilet in the downstairs shower room as I prepare and make food in the kitchen. It is a very crowded party, with sixty people inside. Friends share trays and trays of fancy nibbles, like roasted baby sweetcorn wrapped in streaky bacon or crostini with mushroom pâté. Fiddly things that I have been busy making since dawn. Everyone finds a place to stand or perch on the side of a chair to eat platefuls of home-made beef bourguignon and steaming mashed potatoes with winter greens. We open several bottles of cheap fizz to toast all our helpers and our new neighbours who've been patient throughout all the noisy work. We haven't done any of this on our own as we thought we would at the start. Our lives have become intertwined with those of so many others.

## FANCY NIBBLES

### HONEY MUSTARD SAUSAGES

Prep 5 minutes

Cook 15 minutes

**Ingredients**

good-quality pork chipolatas (2–3 per person)

Dijon or wholegrain mustard

honey

**Method**

1. Place the chipolatas either in a frying pan on the hob with a little oil over a medium heat or in a roasting tin and into a 200°C/fan 180°C/gas 6 oven for 10 minutes until browned.
2. Add 2 tsp Dijon or wholegrain mustard and honey for every 12 sausages and stir.
3. Finish the sausages off in the roasting oven or over the hob for a further 5 minutes or until cooked through. These can be served hot or cold.

## BLINIS WITH CRÈME FRAICHE AND SMOKED SALMON

You can buy ready-made blinis and simply warm them on a baking tray, then leave to cool slightly before adding a scrape of crème fraiche or cream cheese and a little piece of smoked salmon and a sprig of dill.

**Prep 15 minutes, plus resting time**
**Cook 25 minutes**

Makes 45 mini-blinis

### Ingredients

170g/6oz plain flour (or half wholemeal and half plain)
1 tsp fast-action yeast or half a 7g sachet
a pinch of salt
1 egg
250ml/8fl oz lukewarm milk
oil or butter for frying

### Method

1. Make a batter by mixing the flour, yeast and salt in a bowl. Make a well in the centre, crack the egg in and whisk gently, adding a little milk bit by bit until it is all combined.
2. Cover and leave to stand for at least 1 hour. You can make this batter 24 hours ahead and store it in the fridge until ready to cook.
3. Heat a little oil or butter in a wide frying pan and add small spoons – about 1 heaped tsp for canapés – of the batter to make little pancakes.
4. Cook for 1–2 minutes on each side (turn as bubbles start to rise on the surface of the blinis) over a medium heat. Repeat until you have used all the batter.
5. Leave to cool slightly before topping and serving. Or leave to go completely cold and store in a tin for a day or two, or in the freezer, until you need them.

Prep 5 minutes

Cook 40 minutes

Makes 1 dish (double up as you need – also freezes well)

## Ingredients

1 aubergine

2 cloves of garlic, bashed and left in skins

4 tbsp olive oil

½ lemon, juiced

2 tbsp tahini paste

pitta, crostini or raw veg to serve

## Method

1. Heat the oven to 180°C/fan 160°C/gas 4.

2. Halve the aubergine and place skin side down on a baking tray with the garlic cloves. Drizzle with 1 tbsp of the olive oil and season well with salt and pepper. Roast for 30–40 minutes.

3. Allow the aubergine to cool slightly then scoop out the roasted aubergine into a blender, with the roasted garlic cloves (squeezed from their skins), lemon juice, tahini and the rest of the olive oil. Blend.

4. To get a nice dipping consistency, you may need to add slightly more lemon juice, olive oil or a little water, but taste and check it first.

5. Serve with toasted pitta slices, crostini or raw vegetables such as peppers, celery, carrot and cucumber.

SEE P.313–4 FOR **SMOKED SALMON TRIANGLES,
HUMMUS WITH PARMA HAM, CROSTINI, CHICKEN
LIVER PÂTÉ** AND **SMOKED SALMON PÂTÉ**

I hear a gentle knock at the door. James hasn't heard it, but I nudge him to get up from the sofa and whisper that I think there's someone outside. There is supposed to be a light that comes on when people approach the door, but the sensor must be broken. He turns the light on with the switch and opens the door. We don't usually get visitors in the evening. A woman is standing in the shadow of the doorway, and I hear her talking without catching the individual words. I tuck my new baby into her blanket and go from the living room to the door. The woman's face is warmed by the light as she leans in. She is holding a baking tray covered with a tea towel. She passes it to James.

'I heard you had a little girl,' she says. 'I don't want to intrude, but I baked some lemon cake to keep you going. Congratulations!'

'Thank you,' we both say, surprised that the news has travelled this fast and that we are being offered such a perfect gift.

James takes the tray to the kitchen. I can hear little feet on the stairs: Molly hasn't gone to sleep yet and has got out of bed to see who it is. I can smell the warm buttery cake, fresh from the oven. The woman is smiling at my little bundle, and I let her peek into the blankets.

'Come in,' I offer, but she keeps to the door and says, 'I don't want to disturb you, but let me know if you need anything. I'm a nurse, and I live down at the end of the village.'

I tell her that I gave birth at home yesterday.

'How wonderful!' she replies.

'I'll pop the tray back in a day or two,' I say.

She wishes us goodnight and, as I turn to take Molly's hand and lead her upstairs, a tear rolls down my face.

'How lovely,' I say to Molly. 'That lady doesn't know us, but she baked us a cake.'


## Vinegar

Two days earlier, on a crisp mid-October morning, the air is cool and fresh, frost sparkling on the grass, when I let Bramble out at first light. I come back into the house and she climbs back in her dog bed, circling around and tucking her head back into her body and closing her eyes. She's telling me it's too early for her to be up. I pace around the house, back and forth between the three rooms downstairs, picking up toys, straightening cushions and putting books back into the bookcase. I find a soft cloth under the kitchen sink, dab it with vinegar and rub it over the mirrors to clean the fly spots off, as my grandma taught me. I start to make another cup of tea and am leaning forward over the back of a chair in the kitchen as the kettle boils. I can feel my body pulsing downwards and cramps come over me in waves across my lower belly. By the time Molly wakes, I've already put two loads of washing through, folded a basket of clothes fresh out of the dryer and put them all away. I make Molly some toast and I can feel my body changing, like an amber light has been on for a while and now it has turned green. I put a jar of jam on the table and go to find James, pausing on the stairs to steady myself on the banister. My body knows that it is time.

The sun gently warms Molly's bedroom. Diamond shapes of light are cast by the little window frames onto the cream-painted wall. A blackbird sits on the windowsill, watching over the garden, and flutters away when the midwife walks in sometime in the mid-morning.

She finds me standing, leaning on the foot of the curved frame of the cast-iron bed, rocking and swaying between contractions. As each one comes, I close my eyes and shrink out of myself, like I'm heading deep into a forest on my own, no one around me; it feels dark and threatens to engulf me, branches in my way, spiky, and grasping to trip me up. In the back of my mind I know I have been here before, and I know I will reach the other side soon. As I drift deep into my dreamlike state, I hear a voice: 'Helen, are you OK?' A clearing in the forest opens up. I blink my eyes into the light and I am here again, in the room, breathing. I focus on the bed first, the teddies set up neatly from when I made it earlier. Above the bed is a painting of a hare, its large glassy eye softly watching over me.

I am wearing the pale-blue dressing gown James bought me last Christmas. It is velvety and soft against my skin. The cord has come loose through the last contraction, and as I fasten it over my huge bump I feel safe again, cocooned. I pause for a moment.

My midwife offers to examine me, but I shake my head; there is no way I want to move from this place. My feet are planted like tree trunks, with roots growing deeper as each contraction comes. I hold strong to this place. I don't want to be messed about with, to have to move around, to lie down and spread my legs apart to be told how many centimetres I am dilated. If it's only three or four I know I'll feel deflated, and if it's eight or nine I'll know it's nearly time and may panic, because I know I have to find strength from nowhere to push. My body is doing what it needs to do. I have lost track of time. I have no idea how long I've been here. Minutes? Hours? It doesn't matter.

The midwife straps a monitor around my enormous belly, and she listens in to the baby's heartbeat. 'Gosh, that's strong,' she says as she takes her stethoscope away and unstraps the monitor. 'Baby is doing tremendous in there and he or she is exactly where they should be.' I am surprised by how low her probe is on my belly. The heartbeat at my regular scans has been much higher up. Relief pumps through me. I smile and look at her kind face. I trust her completely.

'Do you need anything?' James whispers to the midwife as she sits on the carpet next to me, taking out her folder of endless paperwork. She replies, 'Midwives are fuelled by tea', and he goes downstairs to make her a cup. I hear the sound of hooves clip-clopping past the house. Our neighbour is taking her two horses down the village to their field for the day.

I sway, rock and breathe out through each surge through my body. These contractions are getting stronger and closer together. I refuse the gas and air we have on standby. I snap at James when he suggests getting it from the next bedroom. I know he is only trying to help, but I knew when the canister was delivered and I had to sign a form to store it that I wouldn't want it. I realise he must feel pretty useless right now – all he wants to do is make it better for me. Knowing that it might make me feel light-headed or nauseous is enough for me to refuse it. I really don't want to be sick. Besides, I want to be present for every second of this. Secretly I want to see how tough I am. All I have to focus on is breathing. In and out. Slowly. Deeply and calmly.

As the next contraction starts to take hold of me, I close my eyes and disappear from the room again into myself. I breathe in through my nose as if I'm about to dive deep down underwater, and I take in the longest breath as if I need to stay there for some time. I forget about the world above. I am on my own and I give in, surrendering my mind to my body.

Between contractions I draw my strength by thinking about countless generations of women who have birthed babies in fields, in bare-stone-floor cottages, in woods and caves. Babies born in harsh conditions and in far-away places, far from medical doctors, interventions and hospitals. From the moment I found out I was pregnant with our second child I wanted a home birth. I didn't like hospital that much when I had Molly. They looked after me per-fectly well but I had two midwives on different shifts that I hadn't met before, and James had to leave me soon after she was born. As a first-time mum at that point, I had no idea about the highs and lows my body and mind would go through during the birth and over those first few precious hours afterwards. Everything was unfamiliar; I remember feeling on edge, anxious and uneasy. This time I wanted to own my experience, to be in my own home and have my people around me, so that, when the time came to lose control and push, I'd feel safe.

Now I am surrounded by people who will help me if anything goes wrong. I am not afraid. I have been around animals birthing their young all my life and I trust in the natural process it is. The midwife beside me is calm and quiet. I know she has the same values as I do. We have had several conversations in the weeks leading up to today. My body has done this once before with no complications, I am young, fit and well – there should be no reason to worry.

I hear voices outside. James shouts, 'Have fun with Grandma at the park.' He shuts the car door and the engine starts up with a low whir. A tear rolls down my cheek.

'What's wrong?' the midwife asks, and I tell her I am thinking about Molly.

'I didn't get to say bye. Is she OK? Did she take a coat?' I don't like being apart from her. I don't want her to feel left out. 'Has she got a hat? I love her so much,' I blurt out, and more tears fall. 'What if this baby gets in the way of that – how can there be enough love?'

A beaming smile spreads across her whole face and she chuckles. 'You'll see,' she says reassuringly. 'There's always enough love.'

James brings me a glass of fresh orange juice, cold from the fridge. I take a sip. It tastes tangy and delicious. 'You need the sugar,' he says, and I gulp the rest down. I am tiring but determined to stay standing. Molly was born like this. I want gravity to help me. My midwife suggests I take a little walk to the bathroom, so I hold James's arm and he supports me as I step along the soft carpet, down into the bathroom at the end of the landing. There is a butterfly resting on the latch of the frosted-glass window. James opens it gently, careful not to disturb its delicate wings, and lets it fly away. He leaves me for a moment, shutting the door over a little. The house is so peaceful. The white bath gleams, because I cleaned it yesterday. A basket of fresh towels sits on the tiled floor and an enamel jug full of dried lavender is on the windowsill. Molly's rubber ducks are lined up along the shelf of the tongue-and-groove panelling. Home is exactly where I want to be right now. I crouch down to sit on the toilet, my legs feeling grateful to have the weight taken off for a moment before the next contraction.

All of a sudden my body begins to push down with such force it takes me by surprise. 'I'm pushing!' I call out loudly, breaking the silence of the house. In a second they are all here, two midwives now and James crammed into our little bathroom. James is instantly by my side at the washbasin. I am standing again. I put my arms around him and bury my head into his chest, and he holds me firmly. I feel like I'm cracking open. I feel burning, it is burning down there. I want it to stop. I push again with all my might. 'The baby's head is out,' I hear them say. I keep my chin on my chest and grip James harder, holding my breath as the next contraction comes. I want it to stop and I want to escape my body, but I can't, this is it. They cheer me on with high-pitched excited laughter, and I force every ounce of strength I have to push. I'm not laughing; I don't feel like I can do it, yet I am doing it. I close my eyes. I have to go with it – I know it will end soon but it is crushing me, like it might overcome me. My waters gush out as my baby is born. I feel an immense and instant relief. I was pushing against the birth sac of fluid, the midwife tells me. I'm glad we are in the bathroom. 'It's a girl,' they say.

Moments after she is out of me and in the midwife's hands, she makes her first cry. I hear my mother-in-law open the kitchen door beneath the bathroom and little Molly is chattering away.

'What's that sound, Grandma?'

'That's the sound of your baby brother or sister.'

And then there is a huge squeal of excitement.

I stand at our bathroom sink weeping with joy. My legs are juddering, shaking. I have been standing up for what seems like an eternity. I can feel a gentle breeze from the bathroom window as I clean my legs with a cloth. The midwives are busy wrapping my pink wriggling baby in a blanket that one of them has fetched from the crib. She is scowling at the light and the faces around her.

'I don't ever have to do that again,' I repeat three times to James, like a mantra.

He is laughing, his eyes proud and teary, holding his daughter as if she is a complete surprise to him. 'You've done it, she's here,' he says.

I have done it, I play over and over in my head; I have two healthy little girls just under two years apart. I can't believe my luck. These first few moments are so raw and vivid. Elation sweeps through me and the little bathroom is filled with love and happiness. New life right here in his arms. New life that we made together. New life that I have grown and birthed.

He is holding his baby girl up to his face like a precious gift, kissing her forehead. She is perfect. 'You've done it!' he says again, with more tears in his eyes, and I reply, 'I don't ever have to do that again.'

Her name is Beatrice, I tell the midwife instantly. Bea for short. James looks at me nervously. We haven't discussed this – or, rather, we have, and we didn't agree. He wanted to call her Ruth. 'Yes, Beatrice, that's her name.' I know he isn't going to argue with his wife who has just given birth. 'OK,' he says, and later I hear him struggling to say 'Beatrice' on the phone to his mates, as if it's too posh for him. 'Yeah, we'll call her Bea for short,' he says.

When I look at her face properly for the first time, I know that she is both beautiful and fearless. I walk slowly along the landing of our home and I climb up into our bed. James cradles Bea in his arms by the window and the midwife helps me deliver the placenta swiftly. I barely see it: this incredible thing that has kept my baby alive is scooped up into a bucket and taken away. She tucks some absorbent pads underneath me to protect the sheets and I sit in our bed, propped up by my soft duck-down pillows.

James calls to Molly to come upstairs. She's been busy making ham sandwiches in the kitchen with her grandma. Bramble runs upstairs too with all the excitement. Molly nestles in between us on the bed; baby Bea is tucked into my breast and starting to suck.

Molly watches her, mesmerised, and she wants to hold her. 'In a few minutes,' I whisper to her. The sun pours in through the little window of our bedroom. My whole world is right here around me.

Mum's voice pauses when James speaks to her on the phone. 'Hello,' he says in a serious tone. She isn't really used to him calling her. 'Is everything alright?' She waits nervously, and he announces proudly that she is a granny again. Squeals of congratulations echo into the room. We all laugh at Granny's high-pitched voice saying, 'Oh my goodness! Oh, my goodness!' and I can picture her dancing around her kitchen. We deliberately didn't tell Mum that I had gone into labour as I know she would only have worried herself sick.

I hear her car arrive an hour or so later, then excited voices downstairs as she greets Bramble with much more fuss than normal. She tiptoes upstairs, but I know she is coming. She peers in quietly around the bedroom door. 'Come in,' I say. 'It's OK, she's awake.' And as I show her my baby, her face is glowing with happiness.

She looks down at the little face and says quietly, holding back her tears, 'Welcome to the family', and instantly asks Molly: 'Who is this little person?'

Molly tells her, 'Baby Bea, Granny, this is my baby Bea.'

Mum sets to work in the kitchen soon afterwards. I haven't eaten more than a slice or two of toast all day and I'm starving. She cooks steak and chips for us all, fetching mine upstairs set out on a tray like I'm royalty. Home-made chips, peas and two different kinds of sauce, thick slices of buttered bread on a side plate, and another large glass of freshly squeezed orange juice to boost my vitamin C levels. She says the iron in the red meat will help me too. As I eat the rich slices of steak that she has already cut up for me, she nurses baby Bea, telling her and Molly an animated story of how she woke in the night to the sound of an owl. She went to look out of the window to see if she could see it in the tree, but it made her jump as she opened the curtains because it was sitting on her window ledge. It moved its head right around to stare at her.

She says the owl told her last night that a new person was coming into the family very soon. It was a sign, she says. 'Twit-twoo,' she adds.

## STEAK AND HOME-MADE CHIPS

Choose the best steak you can afford from a farm or shop you trust. Rib-eye, sirloin, rump, tomahawk, fillet or minute/flatiron steaks are some options and they vary in price and size. Timings given here are for a sirloin steak but a larger bone-in steak will need searing as below and cooking in a hot oven depending on its size. You can add mushrooms, onion, garlic and a little cream and Dijon mustard to the meat juices in the pan to make a quick sauce to serve with the steak.

**Prep 5 minutes**

**Cook 15–20 minutes (plus 30 minutes for oven-cooked chips)**

Serves 2

### Ingredients

2 x 1-inch-thick sirloin steaks, roughly 250g/9oz each
butter for frying

### Method

1. Season your steak well then let it get to room temperature before you cook it (never cook it straight from the fridge). Heat a heavy-bottomed pan on a high heat and add a little oil or fat.
2. If cooking a sirloin, which has a good bit of fat under the skin, use tongs to hold the steak with the skin side down in the hot pan first, to sear the fat and release some of it into the pan.
3. Sear one side of the steak, then turn it and baste it regularly, adding a knob of butter during cooking.
4. Cook for as long or as little as you like to get the steak to your preference – rare, medium or well done. This just takes practice and experience. Approximate cooking times for a 1-inch-thick sirloin steak: 1.5–2 minutes per side for rare, 2–3 minutes per side for

medium rare, 3–4 minutes for medium to well done, and 5–6 mins per side for well done.

5. Before serving, rest your steak for up to 10 minutes on a warm plate or board with a rim that can collect the juices.

## HOME-MADE CHIPS

These can be fried or alternatively made in the oven, which uses less fat. I am trying to cut out oils like sunflower, canola, corn and other seed oils from my family's diet because they are highly processed.

### FRIED CHIPS

**Prep 10 minutes / Stand 1 hour**
**Cook 10 minutes per batch**

**Ingredients**

2–3 large floury potatoes such as King Edward, about 700g/1.5lb
(allow 1–2 large potatoes per person)
enough beef dripping for filling a chip pan by ⅓ or a few tbsp for an
air fryer

**Method**

1. Peel, slice and cut the potatoes into the chip size of your preference and leave to stand for an hour in cold water.
2. Heat up beef dripping in a chip pan or fryer.
3. Dry the chips with a tea towel or kitchen paper before frying.
4. Fry the chips in one batch first, until cooked through but not browned, for 5–10 minutes. Lift them out of the pan and let the fat heat up again.
5. Give the chips a second fry until brown and crispy – watch them if you've cut them very thin – then lift out of your fryer, season to taste and serve.

### OVEN-COOKED CHIPS

**Ingredients**

2–3 large potatoes, about 700g/1.5lb (as above)
3 tbsp beef dripping, lard or goose fat

**Method**

1. Heat the oven to 200°C/fan 180°C/gas 6.
2. Peel and slice the potatoes as before, soak in cold water for an hour or so, then drain and dry them.
3. Melt a few tbsp of fat in a large baking tray in the oven. Remove from the oven and toss the potatoes on the tray to cover in the fat.
4. Roast for 30–40 minutes, depending on their thickness, until brown and crispy, and season to taste.

Note: You can leave the skins of the potatoes on if they are washed. I also make tray-roasted cubes of potatoes by adding a chopped onion, garlic and herbs before roasting. You can experiment to find the way your family likes them best.

Mum's mum died a couple of weeks after Bea was born. Mum had visited her days before, and told her our news. She had said, 'Beatrice, what a beautiful name.' Mum and dad went to the funeral; it wasn't something they thought my brother, sister and I should be involved in. When they cleaned my Gran's house out, mum got the white porcelain dog. Her mum had wanted her to have it.

## Trapped

It is Monday morning. I look out of our bedroom window at the sky and see dark clouds looming in the distance. It hasn't really got light yet. The wind is blowing the bare tree branches around by the bus stop; one has cracked and fallen on the grass through the night. Helen will be here soon, and we're going for our regular morning walk and catch-up. I need to get the girls dressed and ready.

I want to attempt to feel slim and in control. I pull on my favourite jeans, long unworn, but they still don't fit. I wriggle out of them and grab the maternity ones I've been living in for months. I had hoped to throw all my maternity clothes in the recycling bin by now, six months after giving birth, but it hasn't happened. Helen went to a mums' exercise group to get fit after having her baby, and has tried to get me to go with her, but there is always a reason I can't make it. James is never around for me to leave the girls with him. He has just started lambing at the farm with his dad. Walking slowly up the road to the next village pushing the double buggy with a two-year-old and a baby and the dog every day is as much exercise as I can face anyway.

Bea is pulling Molly's hair and they are squealing in their bedroom. I rush through, pulling an old T-shirt on, and gently tell Bea to let go and explain to Molly that she is too little to know that she's hurting her and that she just likes her pretty hair. Their bedroom is a tip, with teddies, toys and clothes strewn everywhere. I sift through the mess to find something clean for them both, and tickle Bea to get her to lie down and have her nappy changed, pulling the mat out from under her cot as I catch her. She rolls away from me and then sticks her fingers in the thick white cream I've just opened to put on her and wipes it on the carpet. Molly runs downstairs half-dressed at the sound of Helen arriving with her little girl, Grace. I shout down to put the kettle on, I won't be long. I wipe the cream up as best I can and dress Bea and grab Molly's trousers, scooping Bea up – I hoist her over my arm like a sack of potatoes – and gathering the dirty nappy and a jumper for myself before going downstairs.

I put Bea in her bouncy chair to sit with the other two girls. Helen is wiping the worktop and washing the cups in the sink. I wash my hands, wrap the bread up, put the butter back in the fridge and pick up the girls' breakfast bowls from the high chair and table. Bramble wags her tail around us, wanting our attention, sensing a walk is imminent. Helen chats to me about her mum as I empty the dishwasher. I avoid eye contact with her and check the girls are OK; they're all busy amongst the basket of toys in the next room. Bea is smiling, sitting in the middle of them chewing on a toy Molly has given her.

'How are you?' Helen says, catching a look at my face, which gives it all away in a flash. 'Oh, come here,' she says, hugging me, and I let it all pour out.

I am soon a snotty mess, crying into her shoulder. I get a piece of kitchen roll and blow my nose. 'I'm just so grateful you're doing that,' I sniff. My shoulders drop and I feel a deep sense of letting go all I've been holding tightly together. 'I'm exhausted,' I say. And as she pours the boiling water into two mugs with teabags in, I tell her that James is really miserable in his job, and that we had another huge row last night about money. He slept in the spare room and didn't speak to me this morning.

'Oh, love,' she says, and gives me another hug. 'It's OK. Come on, we'll go and call on Nicola and get out for some fresh air.'

Nicola lives at the top of the village and her little girl, Maia, is a similar age to Molly, nearly two. She is pregnant with her second child. We met through walking the dogs. Her little dog Archie, a cheeky Border terrier, greets us at the door with his friendly yapping. Some gentle music is on and Maia is sitting in the living room playing with her Fisher-Price bus, putting the people in and taking them out. I call through to Nicola, who's out the back putting some recycling in their shed, to tell her we are here. We set off down the road, three buggies, mine as wide as a bus with my two girls strapped in side by side, and two dogs with us. We don't hesitate, because it will only be a matter of minutes before one or more of the children wants to clamber out and toddle off in another direction. We try to get together like this every week, sometimes with another friend we call in on in the next village, but she's gone back to work so it's just the three of us now.

As we walk, we air our problems. We talk about our lack of sleep, comparing our bad nights and how awful these two-year-old girls' tantrums are getting. Nicola complains of having horrid migraines over the weekend and having to get her mum to come and stay over. Helen makes us laugh telling us about her husband being useless at DIY, and proudly tells us she pulled down the old garden shed by herself yesterday, burned it, and started building the new one. I soak up the chatter as we push our buggies up and down the hills, all the time thinking about the resentful words that lashed back and forth between me and James last night. We are hurting each other. He left for work with a face like thunder.

The wind has blown the rain clouds away and there are glimpses of blue sky; the sun is starting to come out. Bramble trots along obediently next to me until she hears the cluck-cluck of a pheasant and I let her race off into the woods to investigate. After several minutes the bird flaps out of the trees and across into the next field. Bramble appears further down the road, padding back to me, wagging her tail, happy that she did a great job.

We walk and talk for the regular three-mile route we take, and plan what we are going to do for the rest of the week. Nicola and I run the local playgroup, and she suggests that she fetches some seeds, pots and compost for the little ones. I say I'll get some new paints as the old ones are dry. Playgroup is in the village hall every Wednesday morning and it mostly consists of us both getting there early to set up an old climbing frame on crash mats and then getting a lot of tubs of well-used toys and trikes out for the toddlers. We drink tea, put out biscuits and snacks for the kids, and catch up with the other mums, keeping an eye on our little people. We try to stop them hitting or biting each other or stealing toys. We set out an activity for the children, but it's usually the mums who end up making the craft, or colouring in the pictures, seated around the little table and chairs while the kids run wild around the hall.

When we get back to my house I pull out a tub of home-made soup from the fridge and Nicola cuts up some bread into chunks. They laugh that I've made roasted butternut squash and carrot again. Helen goes to her car to get a tin of brownies she has baked. We share the food around the table, feeding the little ones first. Nicola walks home and Helen leaves, hugging me and telling me I'll be OK. She's sure everything will be alright by tonight and that she will see me next week; I am to ring her anytime. They don't know how much of a lifeline they are to me.

## LENTIL AND TOMATO STORECUPBOARD SOUP

Prep 10 minutes

Cook 40 minutes

Serves 4

### Ingredients

2 tbsp olive oil

1 onion, diced

1 carrot, diced

100g/4oz red split lentils

½ tsp ground cumin

½ tsp ground coriander

salt and pepper

1 x 400g tin of tomatoes

700ml/1¼ pints chicken or vegetable stock

75g/3oz bacon or pancetta, cubed

### Method

1. Heat the olive oil in a large heavy-bottomed pan and then add the onion and carrot. Sauté on a low heat for about 10 minutes, until soft.
2. Add the lentils and spices and a good grind of salt and pepper. Cook for a minute or two to release their flavours.
3. Tip in the tinned tomatoes and most of the stock. Stir and simmer on a low heat, partially covered with a lid, for 30 minutes. If it looks too thick, add some more stock.
4. Meanwhile, fry the bacon or pancetta cubes and set aside.
5. Blend the soup with a stick blender (or in a blender).
6. Serve with the pan-fried bacon or pancetta cubes and a drizzle of the fat from the pan.

# FIELD MUSHROOM SOUP

Prep 10 minutes

Cook 30 minutes

Serves 2–4

## Ingredients

50g/2oz unsalted butter

2 medium onions, chopped

2 garlic cloves, crushed

500g/1lb mushrooms, chopped – good flat portobello-type
mushrooms (or a mixture) if you haven't got field ones. Before
chopping, clean soil off with a pastry brush and check for any
critters lurking in them – a good tap usually does this.

1 glass white wine

2 tsp plain flour

200ml/7fl oz chicken stock

200ml/7fl oz whole milk

grating of nutmeg

parsley and single cream, to serve

## Method

1. In a heavy-bottomed pan, soften the onions and garlic in butter for
   about 10 minutes on a low heat.
2. Add the mushrooms and cook for a further 3 minutes, seasoning them
   well. Add the wine, bring to the boil then simmer for 2 minutes.
3. Sprinkle in the flour to thicken the soup and cook for a minute before
   adding the stock.
4. Simmer for about 15 minutes before adding the milk and nutmeg,
   heating through without boiling. Blend to finish. Serve with a swirl of
   cream and a sprinkle of finely chopped parsley.

After unloading and filling the washing machine and hanging the
wet things on a rack to dry, I load the girls into the car to go and

see the health visitor in town for a check-up. Bea falls asleep on the way there, but luckily she is happy and smiley when I wake her. After she is weighed and measured, I talk to the health visitor about how she's doing. Molly plays with the toys on the floor. We chat in general about breastfeeding and weaning, and she tells me what a great job I'm doing, and then she asks how I am.

'What do you do for yourself?' she asks, and my eyes prick with tears again. I try to look away and gather myself, but she patiently waits for me to answer.

'Myself?' I think. 'I'm hardly ever on my own. There isn't any "me time". I'm not even sure there is a "me" any more,' I add quietly. I take a deep breath. 'We're starting to apply for planning permission at my husband's family farm.' The health visitor sits quietly, nodding as I go on. 'It's an old stone barn half a mile from the road up a field, on the side of a fell, with no running water, no electric – it's a total renovation job that will cost a fortune. There are even sheep still living in it!' She laughs at this bit, but it is true. 'We have worked really hard to make our home in a village. I really love it, I have friends around me, I can walk to the local school and nursery and the girls like playing at the park. I don't want to leave. But,' I say, and I pause, and she encourages me to carry on by saying, 'Yes, there is always a but', and I take my time, 'my husband is barely ever home, and when he is he's hungry, tired or has more work to do so he has to go into his office, or he just eats and drives off again to the farm. I want him to spend more time with his little girls.' I look over at them on the floor playing, tears dripping from my eyes, and turn away before Molly can see me. 'He is amazing when he's with them, but they are growing up fast and I worry that he's missing out.' I sniff.

'Do you do anything together, just on your own?' she asks. 'Like go away for a night?'

'No, we don't have any time on our own any more. It's hard to go away because I'm feeding through the night and I wouldn't ask anyone to babysit overnight, and he is constantly stressed about money.'

Hearing myself aloud, it sounds like I'm saying money is more important to him than me and the children, and I know that's not true or fair, because he is only working this hard to make a good life for us. But the whole planning permission thing is adding a ton of stress onto him, and as far as I can tell he is already stretched beyond what I think is reasonable for anyone.

I carry on explaining that it is always a point of contention that only one of us is earning. 'We never find any time to just sit and talk about it rationally together; it just builds up and builds up until I complain that he doesn't know what it's like for me, and he says he doesn't have a choice because I'm not earning anything.'

'How do you feel about that?' she asks.

'It makes me feel worthless,' I say, adding, 'If I go back to work – and I don't have anything in particular to go back to – we'll have nursery fees or a childminder to pay to look after them. I've tried to explain that to him, that it will probably cost more than I can earn.' I also say out loud (defensively) as I sniff, 'I didn't have children for someone else to bring them up.' And I blow my nose into the tattered hanky I have tucked up my sleeve.

She passes me a box of fresh tissues. She is listening without judgement and she sits holding my hands in hers, calmly waiting for me to catch my breath again.

Molly and Bea carry on stacking blocks. Bea's legs kick and her face lights up every time she knocks the tower over, her little feet in Molly's old soft-leather slippers with bold pink flowers on them, and her hands flap up and down for Molly to help her build it again. Molly starts to pull the toy telephone with wobbly eyes around the room; it jangles as it moves. I tell her I had the same one when I was a girl, and I pick up the handset and pretend to make a call.

The health visitor is smiling. She tells me again what wonderful girls they are and that they are a credit to me. She gently suggests that I really should try to let my husband look after them now and again on his own, or perhaps I could ask my family to watch them so I can do something for myself. But I already know the answers to these suggestions. I know James would do his best but all he can offer practically is an hour or so while I go to the dentist or get some shopping. I already ask too much of Mum. I don't want her, or my mother-in-law, to look after my children beyond a little babysitting here and there. They have done their time with small children. Asking them to watch the girls while I go for a walk on my own or have a spa day seems absurd.

Everyone's ideas about a sensible, well-balanced life seem to rest on assumptions that don't apply to us. Everyone else seems to have one job, and some free time; they're not endlessly hustling and juggling all these different things – or perhaps it just feels this way because we are worn down. The good advice people offer all rings true, but it just doesn't seem possible right now.

As I drive home, I glance back at Molly in her car seat nibbling on a packet of raisins and Bea watching her like a hawk, and I think to myself that we've built a kind of trap for ourselves. In trying to create our dream life, we've fallen firmly into old-fashioned gender roles. He is the breadwinner, I am the mum and homemaker. I know we have the same values – that one of us should be at home with the children until they're old enough to go to nursery and school – but it feels like I'm trapped at home as much as James is trapped in his job. He never wanted to be that kind of person. He can't switch to part-time hours and look after the girls so that I can work, because the work I'd be able to do wouldn't earn anywhere near the amount of money we need to pay the bills.

I want James to know that staying at home isn't all it's cracked up to be. That I don't sit around watching daytime telly and chatting to my friends. These long days with a toddler and a baby are draining and thankless. I feel like I'm doing everything wrong when I see other women around me. How do they make it look so easy? I feel more like a frumpy mummy than a 'yummy mummy', whatever that is. Daytime TV is full of made-up women discussing the best beauty cream or the latest parenting advice. Every day I just run between jobs as if I have several pans of milk on the hob at once, each one at a different heat, and I'm desperately trying to stop one of them boiling over. I juggle the girls' activities in between nap times, housework, walking the dog, doing the shopping and paying the bills. Their moods are ever-changing. Motherhood is a treadmill of demands – mental, emotional and physical. I never feel like I finish anything or do anything properly. They seem to cry at me all the time. They cry if their plate or spoon is wrong, if they are hungry, too hot, too cold, too tired, too bored, too overstimulated. I wipe their bottoms, clean their faces, wash their sticky fingers and rub the grubby marks off my jumper with the dishcloth. I strip the beds, scoop the sick up from the pillows, scrub the toilet bowl again and wash the soiled underwear. I smile, sing them lullabies, play cafes and hide-and-seek. I read them stories, wash their paintbrushes, wipe the table again and squeeze the Play-Doh back into its coloured pots.

I am the entertainer, the educator, the nurse, the cook, the cleaner, the safe place, the emotional dumping ground.

My body has lost any shape and strength it once had. My breasts ache or drip with milk, telling me I must sit down and feed again. I am permanently exhausted. My brain is foggy and often I can't remember basic things like when bin day is; all the days just blend into one. I overcompensate by working longer and harder to 'keep up appearances'. The only books I read are stories like *The Gruffalo* and *Where's Spot?* Most of the music I listen to is nursery rhymes in the car. Any creative talent I had has died in the slog of domestic life. I have no energy to think or be anything else right now. I have eroded away. I am not the person my husband married any more.

I'm dull, overweight and boring. I'm irrational, jaded and angry. I whinge at him all the time.

I know James wants more time with the girls, and he doesn't want to be as stressed as he is. It all boils down to money and this life we have built – this chasing of the 'perfect life' is eating us both up. I can't see a way out of it. I wanted it all so badly, and worked so hard to create it, but it feels like it isn't enough. Like the finish line has been moved just as we reached what I thought could be the end. He doesn't think this is the end, though; he thinks this is only halfway to the farm. The resentment we have towards each other is horrid. I resent the freedom he has. He drives off every morning and never has to think about anyone else. He gets to spend large parts of the day on his own. I am often angry that I feel like this.

I have so much work that is my responsibility, yet I am invisible because I don't earn any bloody money. All my efforts feel worthless. I have started to hate this 'beautiful' life.

It makes me miserable to see James so unhappy. It makes him miserable seeing me happy with the girls, because however hard it is behind the scenes I try to make it brilliant for them. We have somehow turned into a stereotypical bickering married couple. We find countless little things to hate each other for, but underneath we both know that we can't stay like this forever. I don't know how to fix this.

I wanted to be a mother more than anything, I chose this.

## Eggs

I call in to the local butchers, Bea fast asleep in her car seat on my aching arm, Molly tagging along beside me. I get her to carry some bread and I choose some gammon steaks and pick up some more milk, but I can't manage much else. I'll cook us gammon, egg and chips tonight, and maybe even a rice pudding, and perhaps he'll be over the row by then.

When James gets home, he kisses the girls on their heads and Molly chatters around him, showing him the drawing she has done. He hugs me and says he's sorry, and can we put last night behind us. I turn to the cooker, trying to hold back my tears again, and crack three eggs into the frying pan. Molly has set the table with a random selection of cutlery and melamine plates and beakers. James lifts Bea from the floor, blowing raspberries onto her tummy, and jiggles her bouncy legs into her high chair. He lines some peas up on her tray and counts them with her and gobbles a few, making her giggle. He butters some bread for Molly, and I can see on his face that his day has been long enough already without any more drama. I put the fried eggs, still runny, onto a plate, and as I lift the chips from their pan and tip them into a dish, I think perhaps I have the better end of the deal by staying home all the time. Maybe it is me who has the freedom. Maybe I trapped him. He only wants to provide for us all. No one tells me what time I have to be somewhere, when I can have a break and what I have to do all day. Even if I'm exhausted from the broken nights and constant demands of a baby and a toddler, I still get to choose how I spend my time and I get to spend it all with my children. I glance back to the table and see Bea's little face, all chubby and smiling at her dad, and I pass Molly the chips, telling her to be careful, to blow on them because they are hot.

## EGGS

Free-range organic eggs are amazing. They are a brilliant source of protein and are full of nutrients. They are a meal in themselves.

### SCRAMBLED EGGS

Melt a generous knob of butter in a frying pan, then add 2 or 3 well-seasoned beaten eggs per person. Use a spatula to gently fold the eggs over a low–medium heat until forming loose curds. The care and patience you give the eggs will be rewarded in silky-smooth deliciousness. Scrambled eggs are my favourite and go well with smoked salmon, toast, sausages, mushrooms and grilled tomatoes.

## OMELETTE

For one omelette, I use 2 to 3 eggs, a knob of butter, a tablespoon of chopped fresh herbs, a handful of grated cheese and any other fillings you like, such as a few cooked mushrooms or a slice of chopped ham. Gently melt the knob of butter in a pan, then pour in the beaten egg. Push the egg to the middle of the pan using a spatula, then swirl the pan so any uncooked egg reaches the edges. Cook over a medium heat until bubbling, and scatter on the fillings. When parts of the top look like they're cooked through, roll the pan towards you and fold over one side of the omelette. Leave for another minute to cook the centre through.

## POACHED EGGS

Crack a very fresh egg into a cup. Bring a pan of water to barely a simmer then swirl the water with a spoon. Tip the egg gently into the water. Turn off the heat and leave for 4 minutes exactly. Scoop out gently with a slotted spoon. Two or more eggs in the same pan will need longer.

## BOILED EGGS

Place your eggs in a pan of cold water and bring to the boil. As soon as the water starts bubbling, turn the heat off and place a lid on the pan – 3 minutes for soft-boiled,

7 minutes for hard-boiled. We prefer our eggs soft and runny to dip toast soldiers into the yolk.

## FRIED EGGS

I have never understood the American versions of fried eggs, sunny-side up or over-easy. I just cook my fried eggs on a skillet or frying pan in a little bacon fat or lard and give them 2–3 minutes, spooning a little of the hot fat over the yolk and the white to help them cook.

## EGGY BREAD (FRENCH TOAST)

Beat 2 or 3 eggs per person in a shallow dish and lay in a piece of bread to soak – stale bread or brioche loaf works well – turning it after a minute. Heat a generous knob of butter in a frying pan until bubbling and put the wet bread straight into the pan. Cook for 2–3 minutes each side to brown. Dust with caster sugar and cinnamon and serve with crispy bacon or fruit and drizzle with maple syrup.

## FRITTATA

**Prep 20 minutes**
**Cook 35 minutes**

Serves 6

### Ingredients

250g/9oz new potatoes (waxy salad-type potatoes such as Charlotte)
1 tbsp olive oil
1 onion, finely sliced
1 large courgette or 2 small ones, sliced thinly
75g/3oz unsalted butter
9 eggs
a handful of chopped mint
salad to serve

### Method

1. Heat the oven to 200°C/fan 180°C/gas 6. Boil the new potatoes in salted water for 8 minutes or until soft when poked with a sharp knife.
2. Meanwhile, heat the oil in a 30cm ovenproof heavy-based frying pan and sauté the onion for 3–4 minutes until soft. Scoop the onion out into

a dish. Add the courgette slices and a bit of the butter to the pan and cook for 3 minutes to soften them, turning them occasionally.

3. After draining the cooked potatoes, slice them into 3–4mm-thick slices.
4. Add the rest of the butter, the cooked potato slices and onion to the courgettes in the pan. Stir the whole mixture together carefully so you don't break down the potatoes, then cook over a medium heat until some of the potatoes have a little colour.
5. In a jug, crack the eggs and beat together, then add salt, pepper and the chopped mint.
6. Pour the eggs over the potato, onion and courgette mixture and shake the pan gently from side to side to make sure the egg has reached the bottom. Don't stir as you don't want scrambled egg.
7. Keep over the heat for 1–2 minutes, being mindful not to burn the bottom of the egg, and then lift carefully into the oven to bake the frittata.
8. Bake for 20 minutes until the top is puffy and golden brown.
9. Leave to cool slightly then run a knife round the edge of the frittata and either tip out onto a plate or leave in the pan. Slice up and serve with salad. I like frittata warm, but it is also good stored in the fridge and served cold.

SEE P.315–6 FOR **OSMAN'S SPICY EGGS** AND **ŁUKASZ' SHAKSHUKA**

## Picnic

I am eighteen years old. James is twenty-one. He wants to show me his grandfather's fell farm. He has to shepherd the flock and check the cattle in the barn. The wind blows the car door nearly off when I open it. We have parked at the sheep pens. The ground looks sodden from yesterday's rain. I carefully step around the car so that I don't wreck my clean boots in the mud. I change into my wellies that are in the back. The clouds are racing across the sky and I shout, 'Hang on, wait for me!' but James can't hear me for the wind. He has already jumped over the gate and is marching up the field with a bag of feed slung over his shoulder. I run to catch up, pulling on a hat I grabbed from the back seat, and try to find my gloves in my pockets once I've zipped my coat up. We are heading to an old stone barn near a stream. We cross the steep field, me almost jogging, trying to keep up with him. A buzzard circles above us, so close that I can see its yellow beak and mottled brown feathers. James is at his happiest outside; he strides out, pointing to things he sees, where a fox has been through the fence. And then to a hare – we watch it bound up the field away from us.

We get to the barn, built from blue Lakeland stone. It is hidden from everyone's view in its own little hollow. It was built around 1890, he tells me, to house cattle and sheep in winter, and to store hay above them in the loft. James opens the big double doors by lifting the wooden bar that sits across both, and five young stirks (yearling bullocks) tumble out, their hooves squelching in the mud as they jostle past each other to go and drink in the beck.

James has done this work all his life. He instinctively knows if any of the animals are ailing, what to do if they need anything and how to move around them. 'Come inside,' he says to me as I wait anxiously around the end of the barn for the cattle to all get safely out of the way. Despite supposedly being a 'farm girl', I'm not very experienced around cows. I never feel completely safe around them – I am small and these bullocks can be playful and rough. I pick my way through the muck – water-filled pockmarks left by cattle hooves – and go into the barn. It is musty, dark and warm. James is throwing some fresh straw down on the half that the cattle live in.

I squeeze past the metal partition gate to the other side, where there is a half-open wooden stable door. The view is across the valley to the fells beyond. It is stunning.

'The lake is just over there,' James says, pointing. He starts climbing up a rickety ladder.

'That doesn't look very safe,' I say, but I follow anyway, stretching and stepping over where a rung is missing.

He pulls my hand as I get to the top, onto a wooden loft, and all of a sudden a clatter and flutter of wings is around us; a tawny owl's wing almost brushes my cheek. The owl flies out the open window and down the hillside from the barn. Cobwebs are hanging from the rafters above me.

'It's a bank barn,' James says. 'The door opens out to the ground level at the back here, see?' And as he releases the large arched wooden door, more light floods in. A fieldmouse scuttles under a bale of hay. 'I want to live here,' he tells me. 'How amazing would it be to wake up here every day?'

I smile, thinking he is joking, but I can see a faraway look in his eyes as he stares at the fells. He is deeply serious.

He throws a couple of bales down to fill the hay rack up and I go outside. He shuts the door behind me, saying, 'I'll meet you at the front.' The mud at the back is worse than the front, thick wet clay. I struggle to get to the grass and one of my wellies gets wedged in the mud; I nearly fall forward as I try to pull my foot out. 'The forecast is good, we'll leave them out today,' he says. The cattle are already down the field in front of the barn, grazing.

James started keeping his own flock of sheep on that farm – Herdwicks, a different breed from the ones his dad kept, so he could make decisions about the breeding lines himself. He worked for his dad on the farm, unpaid, for their keep. There were regular remarks made by my father-in-law about how James shouldn't get his hopes up about farming the land in the future because he was planning on selling up. He always complained that he hated the valley in winter. Every year, as autumn turned into one long hard slog of working outside on cold, wet, dark days, his grumbling got louder. Every spring his resolve to not go through that again was stronger. I couldn't see how it would ever work out. There was no farmhouse with the land any more – it had been sold some fifteen years earlier when James's grandfather died. The estate had to be split between the three children, James's dad and his two aunts. James's mum and dad now owned and farmed this land but had no house on it. But farming was like a drug to James. If he didn't work outside on the land with his sheep, he felt adrift.

Over the next few months we had meetings with an architect, builders, surveyors, planning consultants and the National Park. When other people became involved it all felt very daunting. James sensitively opened up the conversation about succession with his parents.

By now it's been eight years since we started making various planning applications, and they've all been refused. Our applications cause unrest in the valley and letters of objection are raised from names we don't recognise. One guy gets a petition together to try and stop any new development. He gatecrashes a planning meeting on our land, striding across the muddy field in his suit and fancy shoes, just off the train from London, waving papers at us. My father-in-law shouts back that this is a private meeting and he should bugger off. We laugh that he wrecked his Oxford brogues. We know that he only wants to protect 'his view' that he paid well over the odds for, but it still hurts.

It seems to me that James's dream is becoming an impossibility. Making a home at the farm is likely to cost way more than we can afford, and the build will take years to complete. I seem to be the only one who worries about the problems facing us with a move like this. There are going to be more costs than just the financial ones, and I don't feel that James ever takes these into account – or, if he does, his desire to make a life for us at the farm is stronger than anything else. I either have to join in or be left behind. If this next planning application is successful, it will change everything. I'll be facing packing up and leaving my home. I'll be leaving friends and neighbours behind. There'll be no more popping into each other's kitchens for a cup of tea whenever we need a chat, or borrowing some flour or butter if we run out. No more walking to school and meeting friends at the swings for the kids to have a run around. The farm is isolated; I don't know if I'll hate being stuck up there. My father-in-law's words echo through my head regularly: 'Who the hell would want to live up there?'

I also know I'll be the one to take the girls out of their lovely primary school, wave goodbye to their kind teachers and plop them down on a carpet in new uniforms in a new school with unfamiliar faces around them. I'll be the one wiping their tears away at bedtime because they miss their friends. And I'm daunted by the packing, moving, fetching and carrying, the sorting and knowing where everything is. I'll be the one with a baby in my arms throughout it all. I'll have to find us a house to rent if we sell ours before the sheep shed is built – we'll need to do that first because to get any electric on site the plan is to put solar panels on the roof. The whole site will be off-grid. And all of this has to be done before we can even start work on converting the barn into a house. I know the paperwork for everything on its own will be a full-time job, let alone making decisions with the builders, sourcing all the materials and fittings, looking after all the kids and doing all the domestic jobs as usual. We're likely to be living in a caravan with three children for months if not years while we do the build, and I hate that prospect.

James is striving for something I don't always believe in and we are becoming divided. But I also know we have to keep trying. As the months pass, this stage of our lives seems like it might last forever. We buy a tatty old caravan to use as a shelter for James and his dad. When I go to the farm I hang out in it with the children and take tins of biscuits or cake for their coffee or teatime breaks. I try not to let the kids crawl on the filthy floor or pick the wings off the dead flies that are hooked in the caravan's torn net cur-tains. I don't like being here, it isn't my space. I'm no use on the farm – I can't catch sheep, turn them over and trim their feet, or load hay onto the quad bike and drive it up hilly fields with small children to look after, and besides, they don't need me to. There is no running water apart from the stream by the roadside and there is no toilet. It is seventeen miles from our home. There is nothing romantic about this dream.

We try and stay overnight in the caravan a couple of times, but it is draughty and cold, and the fire reeks dangerously of gas when it's on, so I leave it off. Lambing time comes around so fast and we only have a couple of indoor places to put sheep and lambs with any problems, one being the stone barn James wants us to live in. Anytime I am here it is muddy, wet and hard to hike up the field with little ones alongside and strapped to my front. He doesn't have time to stop and see the children properly and I don't have any jobs to do here, so I am a spare part.

One day during this period, when the girls are still small, we pack up some lunch to take to James and his dad. They are making hay. I put the radio on and the windows down. Molly and I sing along to Lily Allen, Bea kicks her legs to the beat.

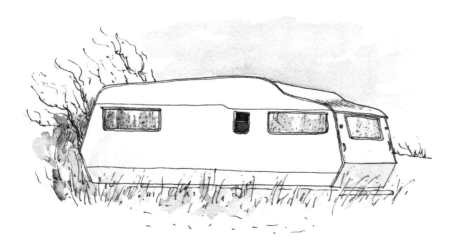

James has been going to the farm every day to turn the hay since it was cut five days ago. He has been watching the weather forecast carefully. His dad phoned early this morning to say the baler was coming and James had to get it rowed up ready.

When we get to the farm I drive through the sheep pens and into the field, where I can see several stacks of bales have been made already. Molly runs off, chasing Bramble. 'Hang on!' I shout after her, and I follow her across the dry stubbly ground and help her climb up on one of the stacks. Bea is now asleep in the car, so I've left the doors and boot open for some light breeze. Bramble is digging for stones in the beck nearby and isn't in the way of the tractors. Molly watches as the baler chunters along behind the tractor, swallowing up a row of dried grass into its belly and spitting out a bale at the back every so often. We count the bales we can see on the ground. James's dad drives another tractor pulling a flat trailer into the field, and James leaps off the back and starts to load the bales onto it. When it is full, they take this load to the stone barn across the road to store for winter.

'Look! Here's Grandad,' I say as he drives towards us. He stops and climbs down from the tractor to give Molly a hug and a kiss. 'We've brought you a picnic,' I announce happily, and he replies, 'That's kind. I've got a sandwich but I won't say no to a bit of cake.' I open the basket and pour him a mug of tea. James has loaded up two stacks while we've been talking. He comes over and takes a swig from the plastic bottle of squash; he is sweating and covered in dust and hayseeds. 'We have sandwiches, Scotch eggs, sausage rolls, apples, fruit loaf and a couple of Mars bars.' Molly is already munching on a ham sandwich.

'We haven't got long,' he says in between wolfing down a sausage roll and unwrapping an already half-melted Mars bar. 'We need to get this field led in and then there's still the Oak Tree Field and the Bridge Field to bale. There's rain coming tonight.' And with that they are back to work, but just before they go I pull my camera from my bag and take a picture of Molly sat atop of the stack of bales, smiling in the sunshine with her grandad's arm around her.

## SAUSAGE ROLLS

**Prep** 30 minutes

**Cook** 40 minutes

Makes 24 small or 10 large sausage rolls

### Ingredients

500g/1lb puff pastry, bought, or 320g all butter and ready-rolled. (You'll have leftover pastry with a block.)

500g/1lb good-quality sausage meat or 6 sausages, deskinned. I always buy our local Cumberland sausage or have meat from our own pigs.

1 small onion or ½ medium onion, chopped very finely (best in a food processor)

small bunch of sage leaves, chopped finely

1 egg, beaten, to glaze

### Method

1. Heat the oven to 200°C/fan 180°C/gas 6. Roll out the pastry into two long rectangles of about 35cm x 11.5cm, roughly 3mm thick, on a lightly floured surface.
2. In a bowl, mix the sausage meat with the onion and sage thoroughly. I find it easiest to mix by hand. You can carry straight on or wash your hands and leave them wet to form long sausages of the meat along the middle of the pastry.
3. Wet along the longest edge of the rectangle with fresh, cold water using a pastry brush or your finger.
4. Clean and dry your hands, then fold over the sausage roll onto the wet line of pastry to form a seal underneath.
5. Place the long sausage roll onto a chopping board and, using a sharp knife, cut to the length you want. Small bite-sized ones are ideal for

children and parties, longer ones for main meals or packed lunches. Freeze raw ones at this point for future baking.

6. Place the cut sausage rolls onto a baking sheet and make a slash or small cut in the top of any large ones to let the steam escape. Brush with the beaten egg and bake for 30–40 minutes, until golden.

When I look at this photograph now, all I see is struggle. There was the constant struggle between me and James about work, money and family life. There was a struggle over how to hold on to this ancient way of life that we were both from. But maybe strongest of all was my personal struggle: my reluctance to throw myself into what was, at that time, his dream.

James has a deep-seated hatred of being 'employed'. It's not the work that bothers him but the being trapped in set hours, having to go to tedious meetings and deal with other people's dramas. So he leaves his fairly well-paid job and goes freelance. It's risky. The mortgage every month is scary; it looms over us. We're dangerously close to not being able to pay it. He rents a little office in the village, but it is too expensive. I help as much as I can, doing some of the admin and research for his projects some mornings while Molly is at nursery and Mum comes to watch Bea. We make a plan to borrow some money and buy a garden office – it will be cheaper in the long run. The wooden office we buy is built behind the house, far enough away that he can work without constant distractions from the girls. I make myself an area in there too, but I barely ever use it. I end up paying the bills by writing cheques at the kitchen table in amongst the Play-Doh.

Working from home was supposed to mean James would see the girls more. That's what we talked about. I foolishly thought he would have structured work-time and home-time, but he never switches off. Whenever I mention this he looks at me like I'm crazy, throws his arms out wide and says, 'Tell me how?' I make him regular cups of tea and carry them out to him from the house. He is in that office day and night, writing reports, filling in funding applications. Whatever pays. He is permanently stressed about getting the next client, to keep the money coming in. If someone asks for something done quickly, he just says yes, even if it means working all night. He can do that. He has also decided to sack the fairly useless planning consultant and is doing the whole farm planning application himself.

I take him a toasted teacake at his desk late afternoon. Helen has been here with Grace. It's sunny and we've been playing in the garden. Molly follows me. 'Come and see the den we've made under the trampoline,' she pleads, pulling on his arm.

I say, 'Sorry you missed your lunch', and ask him light-heartedly if he is nearly finished.

He is sharp to reply. 'Finished? I've got a deadline on Wednesday to get this whole lottery funding thing in. I'm never fucking finished. I'm going to be here all night.'

I usher Molly out, telling her to go and check on Bea, who is asleep in the buggy by the front door. 'I didn't realise the deadline was Wednesday.'

He looks burnt-out and angry. He snaps, 'For fuck's sake, what planet are you living on? I don't get to play teddy bears' picnics in the garden with my friends while someone else pays the bills.'

'Oh, fuck off, I am not putting up with your nastiness. I brought you some food and thought you might just take ten minutes off. We haven't bothered you all day. It was your decision to leave that job. I am doing everything to keep the girls from pestering to come and see "Daddy" every five minutes. Do you think it's easy coping with their tantrums and dramas all the time? I am doing my bloody best.'

'What exactly do you want me to do about it?'

'I am sick of being left on my own with them all of the time.' But he is so angry he doesn't care what I have to say. As far as he is concerned, I'm prepared to sacrifice him to get the lifestyle I want. I'm not taking any of this pressure off him. I don't know how to fix this, and he won't listen to me about ways of changing things. He only knows how to slog it out – that's what he was brought up to do. He is about to have another go at me and I don't want to hear it. Wild with rage at him, I lock the door and shout through the glass: 'You've no right to speak to me like that. I'll throw away the fucking key.' I don't care what he is shouting back at me – he can stew in there.

I go straight to the front door and tell Molly to get her wellies on, then storm up the road to the duck pond, pushing the buggy with Bea still asleep. Molly runs behind me on the grass and when I get to the top of the hill, she tugs on me. 'What's wrong, Mummy?'

'I'm OK,' I sniff, trying not to let her see me upset. I am trying to calm down and don't know if I want to pack both girls up in the car and drive far away from here or fall down in a heap on the grass, sobbing. 'Your dad is tired. He's got a lot of work to do. He didn't mean to say a nasty word. I just wanted him to come and play with us.' I hug her, and Bea sneezes; the sunshine is right in her face. I turn her buggy around. 'Oh, bless you!' I say, and lift her out, cuddling her in close.

James was going to get to the farm by being manic, single-minded and, when necessary, by fighting the world. I was surviving it by digging in and working harder. I got through the tough times by emotionally shutting down. I became cold towards him as a form of punishment, which must have seemed to him like being doubly punished. Our respective roles were making us bitter and angry. We were becoming monsters to each other. The ways we survived at that time hurt each other. It was a mess.

# LATE AFTERNOON

When we get home from school, Isaac gets changed and runs off to find James. He wants to see the new calf. Tom introduces 'Steve the Turtle' to all his other toys. I check the fridge. I have some lamb mince that I defrosted and a few vegetables. I will make the curried lamb and rice dish that I invented a couple of weeks ago. And I will cook a few sausages for the boys, because they like plainer food.

I put the oven on and find an aubergine in the fridge that needs using, so I start by cutting it into cubes. Tom wants to help, so I ask him to put the cubes on a roasting tray. Standing on a stool, he loads them up in the front loader of his toy tractor and drops them onto the tray. I season them with some mixed spice, salt and pepper and a little olive oil. While the oven is heating up, I brown the mince with an onion in my big heavy pan that has a lid. I add half a yellow pepper, chopped quite small, half a courgette and about three mushrooms into the browned meat. I rummage in between the jars on the top shelf of the fridge and find the open korma curry paste. I stir two tablespoons of it into the pan and it sizzles up as I mix it into the juices of the meat and vegetables.

Whatever happens during the day, I do my best to cook an evening meal. Yes, I get fed up of being responsible for what everyone is going to eat, but someone has to be. The worst part is working out what to make. I am not one of those super-organised, meal-planning-spreadsheet people. I'm probably more random, more like my mum, than I care to admit. I get to 5 p.m. most nights not knowing what we are going to eat, just like she did all those years ago.

207

Once I start to make a meal, my irritation vanishes, and I usually feel better as I cook. I take pride in my work, even if the meal is sometimes basic. I have become good at improvising and making a decent meal from a few simple ingredients. I open the fridge to see what needs using up next. I always have a few different cuts of meat defrosting in the drawer and I base my meals around those. We get a local vegetable delivery every week and grow some of our own in the summer, and I keep my pantry fairly well stocked.

Being a farmer's wife has changed my relationship with food. I have always been interested in cooking and baking, but when I was young I didn't think very much about where the ingredients came from. Perhaps I should have cared, given that I grew up on a farm, but for a long time I didn't. I fell into a habit of shopping thoughtlessly, just picking up anything from the supermarket shelves, filling the trolley with whatever I wanted regardless of season or where it came from, quite often choosing the cheapest plastic-wrapped chicken. I didn't properly understand the consequences of my actions. But we now know beyond doubt that cheap food from bad farming wrecks the world.

We need to be highly suspicious of food that seems too cheap to be true, because somewhere a field, an animal, a farmer or a worker is paying the price for that. But a lot of what we eat now is pre-packaged or cooked for us, and these ingredients sneak into our diet in all sorts of ways – in pies and sandwiches, in cafes, restaurants and takeaways, on hospital trays and on school-dinner plates. Doing anything about all this as an individual is complicated. It is not as simple as switching to a 'plant-based diet' to save the planet. The worst farming on earth is acres and acres of wheat, soy and maize grown by ploughing, which creates whole landscapes devoid of nature. These crops are wholly dependent upon synthetic chemicals – pesticides, herbicides and fossil-fuel fertilisers that are disastrous for the soil, rivers, oceans, insects and birds. Eating 'plant-based' products supports these systems. Ultimately, we need to understand ecosystems and farming better to make informed decisions about what to eat.

Not all beef is fed on grain or from a feedlot. It is good to care and want to act to make a change, but I know what a good healthy farm looks like, and it includes grazing livestock. The majority of family farms across the UK care deeply for their livestock and the land they live on. Our cattle and sheep are grown from sunshine, rainfall and grass. They live in a landscape that increasingly, thanks to our efforts, has lots of wild things. Our farm hums with insects and birdsong. We can't produce meat as cheaply as they can in cleared Amazon rainforest, and no one sensible should expect us to. I know we are in the midst of an economic crisis, but for me good food from good farming is the last thing I'd seek to spend less on.

On our farm we try to farm 'regeneratively', which is a fancy way of saying we manage grazing carefully, which in turn improves the soil health, and we restore habitats for nature. We have spent twenty years learning about all this. We believe in being good stewards of the land.

Living on a farm has helped me to see that we all have a responsibility, when we shop, to support good farming. If we don't make this effort, we trap families like mine in bad and unsustainable forms of farming. The industrial farming practices I hate that are damaging to the planet are driven by our insatiable desire for ever-cheaper food. This gets pushed onto all farmers, and then, when it becomes visible, when farmers everywhere try to compete by keeping chickens cooped up, or pigs in dirty concrete pens, or cows that don't even go outside, many people dislike it. If you want something better for animals and the land, then you need to support farmers to do better – and 'better' has a real cost that someone has to pay.

Access to good healthy food is, for me, a basic human right. But it isn't solved by making farming worse in a race to the bottom.

There is also a growing body of evidence that foodstuffs from healthy soil and good farming systems provide better nutrition.

Our shopping, cooking and eating are like votes for better farming and better health. Seeing food as medicine and starting to think about things like our gut microbiome changed my mind about supporting organic food. We are essentially made from what we eat and affected, in turn, by what our food eats.

Yes, doing the right thing is really difficult (sometimes impossible), if you are struggling for money and you aren't an expert on farming, but those of us who can do better – and that is a considerable number of us – have a duty to, and must, make this effort. And there are smart ways to eat well by buying inexpensive cuts of meat, batch-cooking and using up all our leftovers. Not everyone has these skills or knowledge, so we have to find ways to help each other.

I know it is tricky to shop locally and seasonally all the time. I did all the wrong things in the past, and even now, given everything I know, I still find it hard. I try my best not to buy things like plastic punnets of raspberries in December, watermelons in February or asparagus in October. Apart from the fact that they are grown in greenhouses that guzzle electricity, heat and water, they are shipped to our supermarket shelves. And I don't want to eat those foods out of season, because they don't taste right. I am overjoyed to find new-season Jersey Royal potatoes in spring, Scottish blueberries and strawberries in the summer, and fresh English apples and pears on the shelves as they are harvested from our orchards. I don't want to eat an apple that has been stored and shipped in a cold store for months from New Zealand, or chicken in a supermarket sandwich that has possibly come from Thailand.

When we are eating out, I try to support local businesses that buy locally produced meat and serve tasty meals. Smashed avocado on toast isn't for me.

With much more British horticulture and clever use of land like ours, we could grow much more of what we need, and we will be food-secure in these troubling times of war and pandemics.

When I shop now it is really different from how I shopped

before. I am now asking myself, with nearly every product I pick up: Where is this from? How was it farmed? Is it good for my family? The food I trust the most is from our garden and fields. Of course, most people can't grow their own, so finding shops that care about these things for you is really important. Shops have to know that you care about animal welfare, and good farming. I think we all need to make some noise about this and put our money where our mouth is, as best we can. I'm not some kind of eco-saint who only buys perfectly ethical things. I accept lots of compromises. I apply a practical 80:20 rule to the food we eat as a family, knowing the meat and vegetables that make up the main part of our diet are as good as they can be, accepting that the 20 per cent will be food that might be more dubious. Yes, there are sometimes junk foods in our trolley. Sometimes you just want a packet of Jaffa Cakes or a can of Coke.

I add some chicken stock to the pan to stop the mix from drying out and then stir in a packet of ready-cooked tomatoey green lentils. The colours in the pan swirl together; the smell is savoury and warming. The aubergine cubes roast in the oven for about twenty minutes while the lamb sits in the pan with the lid clamped on, letting the flavours settle. I'll mix them in when they are ready. The best pan I have is a heavy Le Creuset that I bought in a sale last year. It is fancy – even at sale price it was expensive – but it is an investment. When I add up how much I must have spent on cheap pans over the years, ones that regularly burned the food, I wish I'd bought one of these at the start. The pan is pale blue, cast-iron, and it keeps the meal warm for ages with the lid on. I use it for so many different dishes.

Tom helps me scoop some rice with a cup from the bag and I show him how to wash it in a sieve under the tap. He tips it into another pan. He wants to stay at the sink and play so I give him a plastic tub and ask him not to use too much washing-up liquid. He fills the sink with warm soapy water, and he runs to fetch a few toy animals that need a 'bath'. I put the rice on to boil and then realise it's time to get the girls from the bus stop in the next village. James sees that I am busy so goes for them.

Isaac comes back into the kitchen and sits on the stool peeling a fresh banana – 'These other bananas have gone a bit brown' – and I ask him if he wants to make his favourite banana bread for pudding. 'I think I've got a bar of dark chocolate hidden somewhere from your dad,' I say. 'Yes, definitely! You need to hide all the chocolate from him.' He grins.

The girls come home and head straight outside because the only thing that they care about right now is the farm. There is the familiar roar of the quad bike engine and a lot of chatter as they go up the hill.

Isaac mashes four overripe bananas. I help him by melting some butter. When he has weighed out the dry ingredients, we stir the butter, bananas and a couple of eggs in his bowl, and then his favourite bit: stirring in the chunks of chocolate I have chopped roughly on a board with a big knife. I hold the bowl up and he scrapes it out into a tin with a spatula. He carefully carries the tin to the oven and puts it in the middle to bake, then turns his trusty cat-and-mouse-shaped timer onto the right setting. He tells me it took longer last time, so he'll set it for fifty minutes.

## ISAAC'S BANANA BREAD

Prep 10 minutes

Cook 40–50 minutes

### Ingredients

150g/5oz unsalted butter, plus extra to serve

4 bananas

2 eggs, beaten

200g/7oz self-raising flour

a pinch of salt

½ tsp bicarbonate of soda

150g/5oz caster sugar

100g/4oz dark chocolate chips or a block chopped into chunks
  (optional)

## You will need

A 450g/1lb loaf tin lined with a loaf tin liner or greaseproof paper

## Method

1. Heat the oven to 190°C/fan 170°C/gas 5. Grease and line the loaf tin. Melt the butter in a pan and leave to cool.

2. In a bowl, mash the bananas and add the beaten eggs, then leave to the side.

3. In a large bowl, sift the flour, salt and bicarbonate of soda. Add the sugar and stir.

4. Stir the melted butter into the mashed banana and egg mix. It's important that the butter isn't hot or it will curdle the egg.

5. Add the wet ingredients to the dry and fold in well. Add the chocolate, if using, and stir.

6. Tip the mixture into the lined loaf tin and bake for 40–50 minutes, covering with foil if the top is browning too fast. The loaf is ready when a skewer comes out clean. Serve warm with extra butter.

The girls jostle through the door, mid-conversation, kicking their boots off, sawdust tumbling out onto the floor. They want food and they want it pretty much now. I quickly wash the baking bowls, because I hate having too much of a mess to sort out after a meal. Isaac sets the plates out and I put the pan of lamb and rice on the table. I slice a baguette for Isaac and Tom, and they will have it with the sausages and some peas. Isaac passes a handful of cutlery across the table and we all grab a knife and fork. Molly fills a jug from the tap, getting the water as cold as she can, and plonks a colourful tower of beakers down. We eat around the table together like this most nights, a family ritual, a coming-together. It feels wrong if someone is missing.

Molly tells us about an altercation with one of her teachers today, about the environmental impacts of different foods. 'They think I'm just in favour of eating meat because I live on a farm.' She is militantly proud of our farm and how sustainable it is, and fierce in her defence of it. Her school recently held a 'Meat-Free Monday'. It provoked a lot of discussion in our area. We live in a pastoral landscape that can't grow crops – we grow grass and farm sheep and cattle. This can be done badly, but at its best it is entirely sustainable. The school should be serving local food from good farms and supporting the local farming community. Bea says that the kids were eating vegan 'sausages' made from 'some sort of processed gunk, and loads got thrown away'. Molly joins in: 'Yes, most folk just had chips, crisps or a cake.' She laughs and says her farming friend Jack told the teacher, 'I wouldn't feed them sausages to a dog.'

I don't mind what other people choose to eat; it is a deeply personal and cultural thing. But whatever your diet, it has to come from sustainable farming systems and that gets forgotten about in binary debates. The plant-based alternatives to meat, in the form of nuggets, burgers and fish-free fingers, are often more harmful to the environment – their primary ingredients come from monocultural crops. Precision-fermented 'protein' grown in industrial vats is not for me – the long list of dubious ingredients, preservatives, flavourings and colours seems to me to be deeply problematic, encouraging people through greenwashing to eat more processed foods.

And I worry about the health crisis we are facing as a nation if we cut red meat and dairy out of our diets. The best meat is nutrient-dense, it fills us up properly. It is food we have thrived on for millennia. The real enemy to our health is processed food. We are addicted to high-salt, high-fat, high-sugar foodstuffs and we have an obesity epidemic. To me, farming in the right way, eating food grown from healthy soil, in the form of simple meat, dairy, fruit and vegetables, with minimal grains, is the answer to healing the earth and our bodies. And I believe we must try to teach our children how to cook and eat proper wholesome food at home and in school. Cooking is every bit as important as maths, English and science. As a parent who buys and cooks the food for my family, I know my role in all of this is vital. I decide what to spend our money on.

I used to think I was a tiny insignificant part in this whole climate conversation, but I have come to realise that through being responsible for my family's meals I have a huge part to play.

I have often felt undervalued, not so much by my own family but in society as a woman choosing to stay at home and cook for her family. There is a strong cultural dogma that life outside the home is more important than the one inside the home. As if domestic work isn't a good enough way to spend a life. But isn't it one of the most important things we can do? Caring for ourselves, our loved ones, our communities and the planet by choosing good food from good farming has been given such low status. I know that we as farmers can do more to celebrate and champion our produce to the public, but it is ordinary people, and still mostly women, like me, the 'mums' in our households, who can make the small changes to the way we buy and cook food that will bring about the big changes that are needed for our planet.

Mums can change the world.

The chatter and news of the day is noisy. Tom stands up on his chair to make himself heard over his sisters. 'I'm a herbivore, like a stegosaurus,' he announces, shoving peas into his mouth. James is onto a second helping by the time I sit down, because I am too busy trying to dig out a pot of natural yogurt from the back of the fridge. The girls and Isaac clear the plates as I finish eating. James makes me a cup of tea. Tom is still at the table twenty minutes later playing with his food, but talk of Isaac's cake being cut up spurs him on to finish.

I might not have felt like cooking an hour ago, but I am reward-ed with more than the food on the table. I know how my children are getting on, what is bothering them or what made them laugh each day. We aren't perfect – we have days when we want to throw the plates at each other. But even on the bad days, having one meal when we eat together is really important to me. In the swirl of the chatter, I learn that Bea got a detention for wearing the wrong kind of socks to school.

# 5

## Birthday Cake

I carry Bea upstairs and she giggles as we pretend to chase Molly, kicking her bum with Bea's feet as we go. They are both covered in sticky icing, sugar and jam, and have grubby hands from holding half-melting chocolate buttons. I am trying to keep them from making handprints on the walls as they jostle each other down the landing. Molly has run off to put a bath bomb in the bath. I call after them: 'Hang on, we need to fill it first and then drop it in gently.' I expect to hear a thud any second as she drops it into the empty tub like a rock, but she has listened this time. They strip off fast in a race to get into the bath; both taps are on full-blast to fill it up. I make them wait as I swish my hands through the bathwater to check the temperature – it is just right. I help them climb in, Molly plops the bomb in and it starts to fizz. They squeal excitedly as they swirl the purple foam around their little white legs as it dissolves. I sit on the lid of the toilet and watch them smearing foam on their heads and chins like beards. They squeal with laughter and offer me a 'beer' from the plastic teacups that bob about in the water, left in from their bath last night. Molly asks me to pass her the toy fish that is supposed to float but always sinks; it lights up and plays music. Bea tries to eat the foam and coughs to spit it out, and then wants me to blow the little bath whistle for her.

The girls and I have just spent the evening decorating Bea's birthday cake. She will be two tomorrow and we're having a party for her. I have been shopping, baking and making up party bags for her cousins and friends for the past few days. I promised them that, if they ate up their supper of beef casserole and mashed pota-toes, they could help me put the icing and sweets on the cake. Two plates of the stew were gobbled up in record time. I cleared and wiped the table and tipped out a carrier bag I had been hiding until that moment – packets of Smarties, Jelly Tots, chocolate buttons, mini-marshmallows and gold and silver cake-decorating balls spread out across the table like a treasure haul. Molly quickly got some little bowls and carefully emptied the packets into them, lin-ing them up ready for us to decorate the cake. Bea pushed her way up onto a chair between Molly and me, nearly tipping the choc-olate buttons on the floor – she managed to grab a few and shove them into her mouth. 'Wait!' I shouted. 'Let's decorate the cake first, and then you can eat what's left.' The temptation to eat them instantly was almost too much for her. I unwrapped some ready-made fondant icing and with both girls kneeling on the chairs at the table I had a captive audience. I gave Bea the pastry brush and a small pan of warmed apricot jam to smear all over the sponge.

## BEEF SHIN STEW WITH ROOT VEGETABLES AND RED WINE

Beef shin is a perfect cut of meat for this inexpensive, delicious winter stew. It cooks slowly and is so tender that it falls apart as you eat it. This stew takes less than 15 minutes to prepare and cooks away all day, filling the kitchen with an amazing smell.

**Prep 15 minutes**
**Cook 4½ hours**

Serves 6–8

### Ingredients
500g/1lb beef shin, diced
2 tbsp plain flour, seasoned well with salt and pepper
1 tsp lard or splash of olive oil (for browning the meat)

2 carrots, cut into chunks

1 parsnip, cut into chunks

1 leek, cut into slices

1 onion, cut into chunks

3 garlic cloves, bashed

1 beetroot, cut into chunks, or a few whole button mushrooms

1 large glass/¼ pint red wine

rosemary, thyme or a bouquet garni (optional)

2 tbsp tomato puree

570ml/1 pint good beef stock

1 tbsp cornflour, mixed to a paste in a little water (optional)

greens and mash, or rice, to serve

## Method

1. Toss the meat in the seasoned flour and heat a little lard or olive oil in a wide frying pan. Place the beef pieces in to brown them, turning when they've got a good deep colour but not stirring.
2. Place the browned meat in a slow cooker or casserole pot.
3. Gently cook the vegetables in the browning pan to soften them. Feel free to substitute the vegetables for whatever you like and have to hand. Turnip, squash or celeriac also work well in this.
4. Tip the vegetables into the slow cooker or casserole and stir into the meat.
5. Pour the red wine into the browning pan, sizzling it up and scraping everything from the pan into the liquid, then tip the wine mixture over the meat and vegetables and add herbs if using.
6. Add the tomato puree and stock and stir well, then put the lid on and set to cook on medium for 4 hours. It is best left to cool and be reheated the next day, as it naturally thickens. But if you are serving it straight away and the gravy is too thin, simmer without the lid on for a further 10 minutes to reduce it, or add the cornflour paste and cook on the hob for a further 10 minutes. Serve with fresh greens like cabbage, beans or broccoli and some buttery mash or sweet potato mash. I also serve this with rice sometimes.

The cake is in the shape of a butterfly. I hired the tin from a local shop and to my amazement it turned out perfectly onto the cooling rack earlier this afternoon. 'When are we doing the cake?' Molly asked me at least a hundred times from the moment I picked her up from school. But I made them wait until after supper because I knew the sweets would spoil their appetite. It worked, although they were getting tired by the time we started. I helped to guide Bea's little hands with the brush, showing her where to dab with more jam, and then I asked them to watch while I gently rolled the icing over the whole thing, covering the cake like a blanket. Molly was quick to get her hands in and copy mine, smoothing, tucking and pressing the icing over the jam and cake. I used a small butter knife to run around the edge and trim the excess icing off, and both girls nibbled at bits of the sweet, pale-pink fondant. Soon we had a pretty but plain pink butterfly cake in front of us, ready to decorate. I explained that the sweets needed sticking down and I made some 'glue' with icing sugar and water and gave them a clean artist's brush each. Bea was more careful than Molly as she started painting and placing the sweets on with my guidance, because she sensed that I might lose my patience with them at any moment. She had her eye firmly on the prize: those leftover sweets. We gently painted and pressed the gold and silver balls down a centre line I scribed with the end of a spoon to make the body of the butterfly, and then started to make swirls and scrolls of Smarties, then circles of glittery coloured Jelly Tots. When finished, it was a work of art.

We are thrilled with our creation. It is going to look wonderful in the centre of the table at Bea's Very Hungry Caterpillar-themed party tomorrow.

I've left the cake on the kitchen table with lots of washing-up still to do, but the girls are really tired and James isn't home yet. I need them in bed so that they can enjoy the party tomorrow. After their bath and lots of naked bouncing on the bed, and a hundred or more requests for more stories, I tuck them both in, asking them to lie quietly and listen for the owls and for their dad coming home. I head downstairs, my body aching for a bath and a lie-down. I'm smiling – we've had such a lovely time together, just the three of us. There may still be a messy kitchen to clean and seventeen layers of pass-the-parcel to wrap up, but that's OK.

Sam, our new young spaniel, runs through from the kitchen to greet me, licking his nose. I can see, instantly, crumbs of sponge on it. As I go down the step into the kitchen, I know what I am going to find. I howl like an animal caught in a trap as I look at the half-eaten cake balancing on the edge of the table. I move it back into the middle. He must have reached up and gulped it from the side, pulling it towards his mouth. Bramble looks at me from her bed by the Rayburn as I take it all in. She flops her head down as I swipe at Sam, who has followed me in, wagging his tail and running around me. She is looking at me as if to say: 'It wasn't me.' I shut Sam out in the yard, kicking the air towards his bum as he races through the door, and shout at him. Then I sit down on the kitchen step, looking at the mess in front of me and the half-eaten cake. It is a disaster.

James's car pulls into the drive, and a moment later he walks in the back door into what looks like a war zone with a sobbing wife. Bramble comes to the door to welcome him with trepidation, but I pat her and she licks my tears. I am too exhausted to start again – I don't even have enough eggs. James says I should just serve what's left, a butterfly minus a wing, and suggests I could patch up the wound with any spare icing. He then shrinks into his shoes as I glare at him. I compose myself as he goes upstairs to check on the girls and I get on the phone to Mum. I tell her what's happened, and she says unhelpfully that I shouldn't have left it there, but that she'll pick up a cake from the supermarket tomorrow before the party. I tell her I've seen a caterpillar one there recently.

I break the news to the girls in the morning and Molly hugs me, telling me it's OK. Bea starts to cry but I tell her Granny will save the day – she is fetching another cake and it will be a chocolate caterpillar – which stops her tears instantly. We get busy loading up the sandwiches I have made and other goodies into the car and then all head off to the local Jersey ice cream farm to set up the party in their barn. The sun is shining. I have organised a friend to do some face-painting – she paints butterflies and caterpillars on the children's faces and arms. Bea looks as pretty as I have ever seen her, with her glittery decorated face, the blonde ringlets of hair trailing down her back, her cheeky smile and button nose. She wears her favourite purple butterfly top, little cotton jeans and new white T-bar shoes. She is happy to be right in the middle of her family and loves the fun of it all. After the food and drink is put away and the kids have had a run around outside, my mum sets out the seats in a line and we play musical chairs and follow-my-leader. Mum soon has all the kids following her like the Pied Piper and we all dance to Bananarama songs from the old CD player we brought. I have long since forgotten about the butterfly cake. We had the best time making it together.

All eyes are on James and me. We are in a room with a horse-shoe-shaped panel of thirteen men and women from the Lake District Planning Authority. We are midway down the room, both standing up after being seated for the past two hours in uncomfortable hard-plastic seats. I've been fidgeting and needing a wee for ages. James takes a deep breath to speak, but in his pause I grasp his hand with my sweaty palm; I feel sick but I start to speak first. My voice is loud, clear and urgent. I haven't planned to say anything, but the words come straight out of me: 'Thank you for considering our planning application. We currently live seventeen miles away from the farm and have two young children, and one on the way.' I rub my right hand across my large pregnant belly. The room is silent but I can hear more words coming from me. I don't pause – they rattle out of me like rapid fire from a machine gun: 'I don't know what we are going to do if we can't live there. Life is really demanding for us as a family, travelling backwards and forwards to look after our livestock. I don't know how we can carry on living like this. It is breaking us.' I look to James. He seems startled at my speech but can't get a word in. 'Farming is in our blood, it is who we are. We want to carry on this fell farm . . .' They all look a bit startled now, and unnerved at this demented and very intense pregnant woman babbling at them. I am pleading with every inch of my being for them to grant us permission to live on this farm. It is so much more than a set of plans and the map laid out in front of them. I want them to know their decision has real human consequences.

'Thank you' – the chairperson cuts me off nervously. That was it. James doesn't get a chance to say anything. Applicants apparently only get five minutes to speak before their case is voted on.

For the previous hour and a half we sat through someone else's arduous application for a three-metre garden wall to be slightly moved, and after hearing both sides of the application it was declined. We held hands and looked at each other, nervously trying to keep positive, waiting for our turn, but I have been losing hope by the minute. I can feel James next to me preparing himself to walk out of here with his lifelong dream squashed by bureaucracy or lost on a technicality. I think I've messed everything up by speaking aloud. They will surely refuse us now. Ours is a big project, a track up a greenfield – 'a scar on the landscape', as one letter of complaint has stated. A new sheep shed to build and a whole barn conversion for us to live in, services to install, new fences and trees to plant. I am shaking. I don't want to look at James again. He is probably cross that I've said the wrong thing or that he didn't get a chance to speak. But he catches my eye as we sit down, his face beaming with pride. He squeezes my hand.

They start voting. I hold my breath and cross everything. We have done all we can. He has worked for months on the planning application, and we have letters of support and full building plans drawn up. Architect's fees, surveyor's bills, landscape architect's reports, wildlife surveys, tree surveys – it has cost us thousands of pounds. In the next minute or two we will find out if it has all been worthwhile.

The men and women of the board are quiet. Papers are shuffled and folded and passed down to the clerk at the end of the panel. A man stands up and announces: 'Twelve votes for, and one un-decided.' It is unanimous. I can't believe it.

We look at each other, checking to see if we heard it right. James kisses me as he punches the air. Everyone in the room around us remains serious. I feel like we've been wrongly accused of committing a crime and are standing in court and have just been given the 'Not guilty' verdict. We can live at the farm. We may pass Go. They have decided our future right here and now. I look at James: the relief and exhaustion is all mixed up in him.

'We've done it!' he says.

'No, you've done it,' I reply.

And we walk out of the room, hand in hand on our ten-year wedding anniversary.

We take a deep breath together as we step outside. Warm summer air hits us and the realisation starts to sink in that everything is about to change. As I walk with James we laugh together. I put my fears aside and enjoy the moment. His passion for carrying on this old-fashioned hill farm and building a life for us there is infectious.

Something in that room, in that moment, moved me to speak out and claim James's dream as mine. I took a leap of faith, in him, in us. I would give up one house that I loved to build another. My family is my 'home'.

Moving to the farm was going to be a major upheaval. Up to that point there had still been a chance it might not happen. My reluctance had been like a protective shield that I'd built around myself and the girls. I had also been preparing myself to help James if the application was refused. But now that it was really happening I could change my reluctance into excitement. I started thinking about what my days could be like on a farm with acres of fields, woods and streams around us. Treehouses, dens and adventures. Sheep and cows and chickens to care for. Maybe even a pony, in time. The project facing us was epic, a classic episode of *Grand Designs* with the standard laughably small budget, only without the cameras. Living on the farm would change our lives for the better in so many ways. It would mean much more time for the kids with their dad, and I would be there to work alongside James instead of constantly wondering when he would be home.

As I locked the door to our home in the village for the last time, I closed a chapter on all the challenges we had faced together. I vowed to start afresh, to meet our new life with energy and enthusiasm, to give it my everything. I drove away waving goodbye to a house that had been a truly special place, with friends around me that I would really miss. I said goodbye to this part of our life, with all its ups and downs, hoping with every inch of me that the next one would work out.

I am pregnant and permanently tired, we are strapped for cash and I have never needed a holiday more. We are driving to the Northumberland coast to stay in my aunt's caravan. It is parked in a field next to a railway line. The caravan shakes with every passing train, but it is close to the beach. We pull up at her golf club, amongst the big posh cars. I scrabble around the grubby footwell of the car for enough change for the parking meter. The bay of Embleton stretches out in front of us, the ruins of Dunstanburgh Castle to the right, and Seahouses along to the left. We load up with our beach gear and wind our way through the dunes. Bramble races ahead. James carries Bea. Molly runs behind him with her new yellow bucket and spade. I have a picnic bag and windbreaker under my arm. I kick my flip-flops off and feel the warm golden sand between my toes as we make camp. I stand for a moment to take in the view. James whistles on Bramble, and she tumbles through the long grass with a length of dried seaweed like a stick in her mouth. I throw it for her and James sets to work straight away, with Molly helping him to build the most epic sandcastle fort. He has reverted to being ten years old again and is in his element. After a while he strips off and races to the sea with Bramble yapping at him as he jumps into the surf. Molly shouts, 'Can I go in?' and has pulled off her sundress and tossed it onto the sand, following him. She squeals at how chilly the water is, but James picks her up and bobs her up and down in the sea until she is used to it. I carry Bea, as she is unsure of it all, and dip her toes into the salty water. My hair catches the light breeze and I look around. There is hardly anyone else here. It is utterly beautiful – a total escape from our daily chores and stresses. I breathe in the sea air and let the waves lap around my feet. I pass James a towel and hug his wet body. We make a vow, there and then, to take a holiday every summer with the kids, however hard it is to make it happen. We need this time together.

## White Toast

I hobble across a dimly lit hospital room into the bathroom with blood trickling down the inside of my legs and a scratchy towel around my arms, just covering my breasts. The light flickers on automatically as I push the door open. I blink – it's too bright after the near-darkness. I step into the cold white tray of the shower cubicle and turn the taps on. Lukewarm water rinses my legs; my already wet hair drips down my body.

'Do you need any help?' the midwife shouts.

'I'm OK,' I call back. I just want to be on my own for a few minutes and press pause – away from the outside world that wants to celebrate our new baby and everything that comes with that.

I just climbed out of the blood-filled birthing pool a few moments ago and have left James holding our ten-minute-old son while I clean up. My body shivers with elation and exhaustion. Tears start to fall down my face and I sniff, but my nose is blocked. I am glad there is no mirror. I turn the hot tap on full, to make some steam to help clear my breathing. I hear the midwife say how well I did to James and he talks to her softly about me being strong. I feel anything but strong right now. But I know I need to find it in me somewhere. I know that the days and months ahead are the toughest. It will be Christmas Day in four days, and I have two little girls aged four and six waiting excitedly for their new baby brother to come home, and for Father Christmas to fill their stockings.

I try to unpack the toiletries bag I brought with me. I fumble for the little bottles of luxury shampoo and conditioner and lavender shower gel. I can't open them. I tip everything out into the sink. They are too small, too fancy and a waste of time. The warm water washes my body clean, but the blood keeps coming. I pack toilet paper between my legs to soak up the flow, but it doesn't stop. The floor soon looks like a crime scene. I stumble trying to reach for another thin towel. I call out for James. He opens the door and catches my hand as I collapse in a heap. He shouts for help from the midwife. I am losing more blood than I should be. They wrap me in towels and help me to the bed. I feel light-headed and as if I'm in a dream. I can only focus on the clouds passing by the window, the first bit of daylight I have noticed today, as the midwife gives me an injection and raises my legs to check me over.

There are new voices and bodies around me – they tell me to breathe in and out with a mask over my face. I lie still as another nurse stitches me up. They tuck me up in blankets to stop me shivering and tell me to rest. My blurry gaze drops towards the crib beside me and I blink, coming back to myself as the gas wears off. My baby, Isaac, has wriggled out of the blankets James set him down in. His little fist is in his mouth, sucking furiously. I move to try and sit up. 'I want to hold him,' I say. James tells me to stay still and lifts Isaac gently, cradling his head. Confidently and calmly he passes him to my breast. He is still sucking on the knuckle of his thumb and the midwife says that, if I feel like it, I can nurse him. It may help reduce my bleeding. I snuggle his bare body onto my naked belly, and he finds my nipple easily. I think back to how long it took me to get my firstborn, Molly, to latch on, and how worried we were then about everything as first-time parents, and back to the many toe-curling moments I had learning how to feed her. Isaac takes the milk like a lamb and my belly constricts and contracts. I have had a haemorrhage, but I don't need a blood transfusion. I am glad I'm here in this hospital room. It isn't home, it isn't where I wanted to be for this birth, but I am really glad I'm here.

It is the shortest day of the year, the winter solstice; it is wild, wet and windy outside. The day started with a trip to our local doctor's surgery. I was eleven days overdue and was having trouble breathing with a bad cold, and feeling really low. We left the girls with my sister-in-law and James took me for a check-up appointment. I had a chest infection and needed antibiotics, but the doctor assured me I could take them whilst pregnant and that I would be having the baby before I finished the course. We went to the local supermarket to pick the prescription up. It was noisy and bustling with people stacking their trolleys high with their big Christmas shop. The lights and music just made me feel worse. I wanted to escape this place and escape my body. I wrapped my coat around me and pulled my hat down, trying to hide. We bumped into some neighbours who wanted to stop and chat. They laughed at me as they said, 'Still in one piece, then.' I tried not to roll my eyes at their attempt at humour – I felt rotten. James was friendly but firm with them as he could see how vulnerable I was feeling.

The doctor had booked me a check-up appointment at the city hospital, and I had my bag in the car just in case I needed to stay. We stocked up on sandwiches and chocolate bars and disgusting energy drinks before driving forty minutes up the motorway, away from home. When I met the midwife who would look after me, I repeated what I had told the doctor: 'I can hardly breathe through my nose I'm so bunged up – how on earth am I going to give birth?' She reassured me, 'Your body can do this. You can do this.' I didn't have much faith in myself any more. I wanted to curl up in bed with a hot lemony drink and sleep.

She left us to settle in to a large, brightly lit room, with the familiar plastic crib made up ready next to the bed. I unpacked a few things and James put Snow Patrol's album on the mini-speaker we'd brought, and he rubbed my back.

The midwife said that as this was my third baby, she could just break my waters and my body would probably kick into action, knowing what to do. I was relieved to hear this as I didn't want the drugs, the drip or any other interventions. I was determined to refuse them if she offered them. I told James that he was to refuse anything on my behalf too. If I could have stayed at home longer I would. But I knew this baby was getting too big inside me and I was really struggling with basic everyday tasks. I reluctantly agreed to her offer – I wanted my baby in my arms. I wanted Christmas Day at home with us all together. I held on to James as she talked to us. Then I welled up and said, 'I'm tired.'

He looked at me and said, 'It's your decision. We can go home if you want, but I know you can do this.'

We have always believed in each other when things get tough.

Three hours later, after my waters had been broken, and after a lot of walking up and down the corridors, pausing as contractions came and holding on to James as if an earthquake was happening inside me, the midwife offered me the birthing pool.

'Anything,' I said desperately. 'Anything.'

I struggled down the corridor, past the nurses' station, ignoring visiting families on the ward and keeping my head down, not catching anyone's eye except for that of a mum who was nursing her newborn daughter. She nodded and smiled a smile that said to me: 'I know what you're going through right now, and you have got this.'

I lowered my heaving body into the warm water as the taps were still filling the pool up. I felt a huge sense of relief and calm wash over me. The taps stopped and all was quiet; the midwife had left us for a few minutes. The water helped me breathe and the room was dark and peaceful. Half an hour later, after swaying and breathing and pushing, Isaac was here. I called out for him to be pulled up out of the water as he was submerged. The midwife assured me he was fine as he was still attached to me by the cord, but I called again, 'Lift him out, help him.' As James helped them bring him up to take his first breath, I looked down. I was a bloody mess.

I stay in hospital that night. They want to keep an eye on me. I want to go home more than anything but have to let them look after me. A lovely fresh-faced young midwife comes in at midnight and takes Isaac off to cuddle him as I steam my face over a bowl of boiling water with lavender drops to try and clear my nose and head. Then I must drop off to sleep in the early hours of the morning. The ache of my breasts stirs me to feed my new baby again, who is now back lying asleep next to me in the plastic crib. As I stir, the midwife instantly comes in to help me lift him out and tuck him in for a feed. She brings me several slices of white toast. I thickly smear them with butter and scrape out the little packets of jelly-like marmalade, and sip my tea. As hard as it has been, I am overcome with love for this magical little person in front of me.

Isaac is pink and healthy and completely brilliant. 'He's a belter,' Cousin Louisa's message says when she replies to James's photo. Hundreds of well-wishes have pinged back onto his phone since he shared a photograph. Isaac has been smiling at me with his eyes from the moment he was born. I am now a mum of three.

When we carry him into the house later that day, his two big sisters take charge and place him in the centre of all their games. I have a couple of days of sleeping and feeding him before we enjoy the last Christmas in our home in the village.

*Trifle*

I walk into my friend's bungalow, the doorway all lit up with fairy lights. I proudly pass her a glass bowl of raspberry and amaretto trifle I have been carefully nursing in the car, and pop back for the panettone pudding that has travelled in a basket in the boot. Her husband takes our coats and dumps them in his daughter's bedroom. 'Wow,' she says. 'These look amazing.' I want to say 'I know' but don't.

The house is noisy and half-full already with their friends and neighbours. I help our girls unzip their new boots and they run off to play on the giant dinosaur in the corner that Santa brought for our friends' son a few days ago. James carries Isaac in his car seat. He is fast asleep and oblivious to the lights, people and music. More people arrive behind us. I follow my friend into the kitchen, hugging other friends on the way. 'You look great,' one of them says to me. I smile back. I don't feel great, that's for sure, but I am here.

# PANETTONE (BREAD AND BUTTER) PUDDING

Prep 20 minutes

Cook 45 minutes

Serves 6–8

## Ingredients

1 (750g) panettone/1 large brioche loaf/8–10 slices stale white bread

50g/2oz unsalted butter, softened, and a little extra for greasing

75g/3oz dried fruit (sultanas, currants, cranberries), dark chocolate
chunks or a few tsp bitter marmalade

300ml/½ pint whole milk

300ml/½ pint double cream

3 eggs

75g/3oz caster sugar

1 tsp vanilla extract or 1 vanilla pod, split and seeds scraped

2 tbsp demerara sugar

Note: If your panettone already has dried fruit or chocolate, you don't
really need to add extra.

## Method

1. Heat the oven to 180°C/fan 160°C/gas 4. Butter a 20 x 30cm baking dish.
2. If using stale bread, butter the slices on both sides. If using panettone
   or brioche, don't use any extra butter, just grease the dish.
3. Slice the edges off the panettone. Line the dish with the slices and tear
   up rough chunks of the rest of the pannetore with your hands for the
   filling, adding fruit or chocolate if using. Stud with 3-4tsp of marmalade
   if you like. If you are using bread, cut four slices diagonally and line the
   dish the triangles then tear up the rest into rough chunks for the filling.
4. In a pan, heat the milk and cream gently until just below boiling. Leave
   to stand for a minute.
5. In a bowl, crack the eggs and whisk with the sugar and vanilla extract
   or seeds.
6. Stir in the hot cream, whisking well to form a custard.

7. Pour the cream mixture over the bread and tap the dish to spread the mixture evenly. Sprinkle the top with demerara sugar.
8. Bake in the oven for 20–30 minutes until golden brown.

The oven is on, and I check the pans of rice simmering. 'I got boil-in-the-bag to make it easy,' my friend says.

I ask if she's OK, and I thank her for hosting the New Year's Eve party. I put the panettone pudding in the oven, and ask if she needs the oven for anything else – no, it's fine. The slow cooker is on the worktop bubbling away with a curry. I say, 'I'll just turn it on really low to keep this pudding warm. What can I do to help?' I add.

'It's fine, have a rest,' she says. 'Have the evening off.'

But the truth is I don't know how to have an evening off. I've become that mum who puts everyone else's needs before her own – there isn't much of 'me' to have a conversation about anything. I'd rather hide in the kitchen and keep myself busy. If I get too comfortable on a sofa I know I will fall asleep. Isaac is ten days old and I feel just being here is an achievement. I feel drab in my maternity top and the same stretchy black trousers I've worn for months. The other women are in glittery dresses and heels, and their lips and eyes sparkle with make-up. Their kids are mostly over five now; some even have teenagers. My hair is tied back in a pony-tail and the only make-up I could find was an old lip gloss, applied in the car on the way. I nearly cancelled coming, but I knew how much the girls would enjoy it. We have just been at home these past few weeks, waiting for this baby to come, and I haven't felt particularly sociable or had any energy to go far. Now he's here I wanted to feel a little more 'normal', so said I was feeling OK to go.

Making a trifle is pretty much an all-day effort. If I'd been really organised, I'd have made the jelly last night and had it ready in the fridge. But despite trying to kid myself that I have it all together like some sort of domestic goddess, I don't really. My days are blending into one tangled blur of breastfeeds, changing nappies, washing, cooking and cleaning up. I barely know what day it is. So it's only at breakfast time this morning that I start snipping the jelly cubes with kitchen scissors into a bowl.

## JELLY FOR ALL OCCASIONS

### REAL FRUIT JELLY

**Prep 15 minutes**

**Cook 5 minutes**

Serves 8

**Ingredients**

1 litre/1¾ pints apple juice

225g/8oz frozen raspberries

1 tbsp gelatin powder or 4 sheets of gelatin soaked in cold water

**Method**

1. In a pan, boil half of the apple juice with the raspberries.
2. Whisk in the gelatin powder or soaked sheets.
3. Stir in the remaining cold juice.
4. Strain through a sieve into a serving dish or glasses, or a greased mould if you're keen to turn the jelly out.
5. Refrigerate overnight until set.

   Note: This can also be made with a good-quality pressed fruit juice. Omit the whole fruit and use half the juice. Boil half, add the gelatin, then add the cold juice before chilling.

### ALCOHOLIC JELLY SHOTS

**Prep 7 minutes**

Makes 16 double shots or 32 singles

**Ingredients**

135g/5oz pack fruit jelly

280ml/½ pint boiling water

180ml/6fl oz chilled water

100ml/4fl oz chilled vodka

**Method**

1. Cut the jelly into cubes and dissolve in the boiling water.
2. Add the cold water and vodka, pour into shot glasses and leave to chill and set overnight.

Molly watches me pour boiling water over the cubes and as I start to stir them together she can see that what I'm doing looks quite good fun and wants a turn. 'Careful, it's hot,' I say. I pass her my fork and watch her copy me, stirring and squashing the little jelly squares onto the side of the glass jug until they melt into the water.

'I want to eat jelly now,' Bea says with a mouthful of Rice Krispies. I explain that we need to add cold water and pour it into the dish over the fruit and let it set first.

'Here, put these into the dish,' I say as I pass her some frozen raspberries, and she dots them around the bottom of the fancy dish carefully. I slice a ready-made Swiss roll (because who has time to bake the sponge?). It's a good one from a local bakery that I've kept hidden from James, because he would have eaten it. I place the circles of sticky jammy cake over the raspberries, giving the girls one half-circle each to eat, and then I hear Isaac cry out on the baby monitor upstairs. I look at the time: he has slept well since his last feed at 4 a.m. – it is nearly eight. We are all still in our pyjamas and dressing gowns. I pop some cold water into Molly's jelly jug, stir it well and pour it carefully through the cracks between the sponge circles, wanting to cover the fruit but not wanting to soak all the cake completely. I pull out a bowl of soup from the fridge to make room, squeeze the dish in beside yogurts and cheese, and carefully close the door. I'll cover it up later.

SEE P.303 FOR A NOTE ON BUYING CAKE

I call for the girls to follow me upstairs to come and get dressed. I scoop Isaac up from his crib next to our bed and climb back in to snuggle with him. I kiss his sweet little face, but I can feel he needs changing before I feed him again. I ask Molly to pass me the changing mat and basket from the floor, and lie him out on our bed, tickling his toes. Bea climbs up to help me.

I spend the rest of the morning cleaning up, hanging wet washing on the rack and putting another load on in the little shed outside the back door where the washing machine is. Isaac is easily entertained in his bouncy chair while the girls play, with CBeebies on in the background. I dash outside to fill the log basket and put three logs into the wood-burner to keep it going for a while. Sam and Bramble circle around me as I go in and out of the kitchen. They have waited long enough – it's nearly 11 a.m.

I bundle everyone up in coats, hats and scarves and manage to find matching wellies that have been tossed in the heap after our last walk. We set off up the road together. I make the girls hold on to the pram until we get to the wider bit of grass. A bumpy walk will help get Isaac back to sleep. He is gazing up at the sky and the jackdaws call all around us. The girls run off up the grass to the duck pond to throw some stale bread crusts in. 'Don't go in the water!' I shout after them. The day is grey and still, mizzling steadily, droplets of rain forming slowly on the branches. I let the dogs off their leads, and Bramble races up towards the girls and fetches a ball back to me that she has found in the long grass. Getting us all outside feels so difficult sometimes. The girls don't ever want to go out, but we have two dogs and they need a walk every morning. I tell them all the time, 'You'll feel better when you come back in, fresh air does you good', like something my mother would say.

When we get back I park Isaac outside by the kitchen window, all snug with the rain cover over him. I have done this with all of my babies, never wanting to disturb their sleep and letting them get as much fresh air around them as possible – but also wanting to grab a few more minutes of peace to get on with my jobs before they wake up.

I get back to making the next stage of the trifle. I put a pan of milk on to boil before I take my own coat off. I measure the custard powder and sugar into a bowl. I use good old-fashioned Bird's custard powder like my mum does, because home-made custard never sets properly. A trifle is all about creating distinct layers; I don't want a sludgy mess. I stir a little cold milk into the powder and sugar to make a paste and then, just as the milk is rising in the pan, I pour it onto the paste and stir well. This will need to cool for a while before being poured over the sponge and jelly, so to stop a skin forming over it I sprinkle the top with caster sugar. I take the cold glass dish out of the fridge. The jelly is still too wobbly – it will need another hour. I decide to pour a little amaretto over the sponge layer because the kids won't really eat this dessert, they'll probably have ice cream and sprinkles. It smells strong and sweet and I only use a tiny bit and then drink the rest of the glass in one. It's like fire down my throat, but the almond aftertaste is so good.

I make my coffee stronger than normal. I'm going to need all the help I can get to make it through today. I look out at Isaac sleeping. I probably have about half an hour to get lunch ready before he'll need another feed. I grate some cheese then cut up some chunks of bread, an apple and a few grapes. The girls want crisps, but we've run out. I find them some crackers and breadsticks. They scatter cheese all over the floor and cracker crumbs all over the table. I brush them off and Bramble cleans them up before I need to get the vacuum out.

James is back home and famished. He's been away since first light. He makes toast and fried eggs and we catch up on what he's been doing. I whip some cream with an electric hand whisk and the noise wakes Isaac; James goes and picks him up out of his buggy and has a cuddle. After another feed and change and settle, I pass Isaac to James, who is nearly asleep on the sofa, and go back to my trifle. The jelly has set, the custard is cool enough to layer up and I feel like I am winning. I top the whole thing with piped whipped cream and decorate it with raspberries and dark chocolate curls that I grate off a block. I might be knackered, but it's a showstopper of a trifle.

I have an extra pair of hands now to help with Isaac and I momentarily feel invincible. I pull my large, patterned ovenproof dish from the back of a cupboard and start to put together a bread-and-butter-style pudding to take along too. Being in the kitchen doing something like this is a break from the nappies, the jangling toys and cleaning up around the house. The girls have run off upstairs to play with their new Sylvanian Families house that Santa brought, and as long as they don't start fighting about how to set it up I think I can get this done. I slice the edges off a large chocolate-chip panettone, which I've also been saving for this, and line the dish with the pieces, crust sides to the dish. I tear the remaining cakey bread into a bowl, eating a lump or two as I go. I could get the girls to help but they would just want to eat it all, and I need to get on. I have another custard to make, but this time a proper one with eggs, sugar, milk and cream whisked together when hot. I pour the steaming mixture over the chunks of panettone and scoop it all into the pre-lined dish. The next job is to add a little of Mum's best marmalade with a teaspoon, poking it into a few gaps I make with another spoon, and I add a few rough chunks of dark chocolate to make it extra-special because I already have a block open from decorating the trifle. I sprinkle crunchy demerara sugar all over the top and bake it in the oven for forty-five minutes until the whole thing is golden brown. Everyone loves a warm spongey pudding and this one is indulgent and delicious. I know that, once broken into, it will be gone in seconds. Turning up to a party with two home-made desserts is my way of looking like I have it together. I would prefer to crawl back into bed, but that would mean admitting I've been defeated by having three kids and

I want to prove to myself (and everyone around me) that I can do this, and do it well.

There is an insane amount of washing-up to do before we are ready to leave the house, but I am determined to leave it tidy, because coming home to an empty sink and clean work surfaces is like winning the lottery to me right now. I'm giving a gift to my future self. I shout instructions to James to feed Bramble, pick up the toys in the living room and draw the curtains. I go and dress the girls in their best party dresses and plait their hair. I pack a nappy bag for every possible scenario, including PJs for the girls and their new slippers to run around in later tonight. We are ready.

After we say our hellos and the girls have run off to play, I fill a tumbler of water from the kitchen tap and take it through to the conservatory/bar, which is a large airy space on the back of the bungalow. The wall behind the bar is mirrored and has glass shelves to showcase my friend's husband's global collection of fancy spirits. He stands proudly behind his bar and serves up all sorts of drinks like he's Tom Cruise from *Cocktail*, and everyone teases him about building a pub in his home. His wife always complains that he's also taken over their bedroom with his exercise bike, set up as if he is racing in the Tour de France.

I go and check on the girls, who have opened up a new game and are playing it on the floor. James is chatting, with a beer in his hand. I notice Isaac wriggle and blink in his car seat. I unpack him like a parcel out of his snowsuit and carry him to the sofa to let him stretch out and wake up. I don't want him to fall back asleep; I know I'll suffer later if he naps too long now, and he's getting too hot. I sit on my own and hear conversation from the conservatory.

'Yes, the baby is here – is that number three?' someone says.

'Yes, she's had another so she doesn't have to get a *real* job,' someone replies.

It is meant to be jokey banter, and if I had the ounce of energy a sense of humour would require I would have shouted through,

'I heard that! Get lost, I have a real job.' But instead I take Isaac to the bathroom, shut the door and lay him carefully on a clean towel on the floor while I take a minute away from everyone else. I run the tap and see my flushed face in the mirror. Tears well in my eyes, but I won't let them fall. I want to leave right now. Why did we even come?

I look at Isaac, ten days old and perfect. I take a deep breath, splash water on my reddened face and carry my tiny baby back into the living room. Another mum asks for a cuddle and I say I need to feed him, wanting to keep him close to me and trying not to crack. I tuck myself into the corner of the sofa and snuggle him under my top. James comes through to ask if I need anything. I tell him to make sure the girls get some food. He can see in my face that something has happened and we won't be staying long. He kisses my head and says I am doing brilliantly. I put my arm around his neck and whisper, 'I love you.' As I sit feeding Isaac, my mind races. I turn my humiliation into anger: I don't have to justify having three children to anyone.

I feed, wind and change Isaac and pass him to a friend so I can get some food. After everyone has finished, I help gather the plates up, then take the pretty blue dish of warm pudding from the oven and put it on the kitchen table. The trifle stands proudly centre stage in the sparkling glass dish that Grandma gave me as a wedding present. I only use it on special occasions. I want to leave, but I will smile through this. Everyone tucks in and the bowls are soon empty. I get lots of compliments and requests for the recipes. My friend says she doesn't know how I do it – 'Ten days after having a baby and turning up with two amazing desserts.' We gather the girls up, much to their outrage at having to leave well before midnight.

I kick off my shoes as we get back home and practically crawl up the stairs, with James ushering the girls up and going back down to fetch Isaac to me so that I can settle him into his crib next to our bed. James goes back to the girls and tells them a made-up story about a mouse that slept through New Year's Eve and woke up in a magical world.

I lie awake, unable to switch off. My mind replays hearing those words. This is enough. I am enough.


## Baguette

All the parent-and-child parking spaces are full. I drive around again, hoping someone will leave. It is starting to rain. I have all three children in the car. The girls are tired from their busy day at school. We are outside the big Morrisons supermarket in our local town. No one is leaving. I drive round and round the car park. Eventually I spot a man loading his shopping into the boot of his car, so I put my indicator on and the windscreen wipers and wait. He takes about three years to return his trolley, get into his highly polished Audi and reverse out. I take about three seconds to fill the space when he pulls away. I leave the kids in the car for a moment while I run for a trolley, scrabbling around for a pound coin in the bottom of my nappy bag to release it from the chain. I squeeze in between my car and the next one and lift Isaac, who is nearly two, into the trolley seat. He is too big for the seat but it's easier than trying to push the trolley and hold his hand as he toddles across the car park. It is a grey Wednesday afternoon in November. I pull his coat around him and put mine on before we troop across the busy car park, picking my way through the rows of cars and puddles. I get us all safely inside the warm, brightly lit store. My damp coat feels thick and heavy.

I switch into Shopping Mode.

I hoist a huge pack of discounted toilet rolls from the pallet in front of us onto the catch under the handle between Isaac's legs. 'Can we get some strawberries?' Molly pleads. 'Go on, then,' I say, and she passes me two punnets. The girls are excited to be out of the car and want to run off, but I keep them close. I pick up three cartons of milk, a large block of cheese and some thinly sliced packaged ham to make the sandwiches Isaac likes. We go up and down the aisles, filling the trolley like this, half to some unwritten adult plan and half to a series of child demands. I am trying to think about what we really need, and all they are seeing is chocolate mini-rolls, jam doughnuts, frozen curly fries and Ben & Jerry's Cookie Dough ice cream. I can't face an argument about every single thing. The girls are restless from being told what to do all day at school. They start shoving each other out the way to get the glory job of pushing the trolley with their little brother in it. He is happy to see them after a day with me and he likes their attention and giggles at them. I reach for James's favourite yogurts and butter, gathering up the tubs and pots in my arms, and then turn around to find the trolley gone. A lady passes me and says crossly under her breath, 'Can't you keep them under control?' The girls have whizzed Isaac along, banging our trolley into hers and shouting at each other. I smile and want to say 'What would you like me to do, take them outside and beat them with a stick?' But I hold my tongue and say, submissively, 'I'm sorry.'

I really don't want to be here.

Another, kinder woman in the next aisle gives me a knowing smile as she sees them laughing and wheeling Isaac around in a circle. I take another deep breath. I see a woman at the end of the aisle who knows my mum, so I turn the kids around and go and hide down the next aisle until I know she has gone the other way. I am not in the mood for small talk.

We have run out of food. I spent the morning in the sheep pens helping James amongst the flock, managing to keep Isaac occupied by giving him little jobs passing me things and stopping him from being splashed by the toxic chemical in the sheep footbath or being trampled and knocked over by the excited sheepdogs on the shitty concrete. Then we walked behind the flock as we moved them to a fresh field. Back at the building site that is our house, our builder, Denis, gave me a list of supplies he needs for the next few days and I called around several suppliers to get the best price, in between making scrambled eggs and toast from bread crusts for us all at lunchtime. I arranged the delivery of a few pallets of breeze-blocks and I talked to Denis about how I want the windows to look. Then I called in a dead sheep to be collected by the knackerman. I haven't answered any of the 'urgent' questions from the insurance company in my emails, and I still need to phone the vet because Bramble is off her regular dog food. In a ten-minute tea break I helped Isaac set up his train track and then it was time to go to the school again. It will be suppertime soon. I don't know what we are going to eat tonight. I hand Isaac a torn-off chunk of baguette to pacify him as he tries to climb out of the trolley.

Bea says she needs the toilet. I am in the cereal aisle, about as far away from the bathrooms as we can be, so I ask her if she can wait, but she can't. I look at our overflowing trolley and the queues for the checkouts, sigh and turn us back towards the entrance to the

store. I leave all my shopping beside the bathroom door, lift Isaac out and try to catch a member of staff to say, 'This is my stuff, we're just in here' – but I don't succeed. I risk it, wondering who would want a trolley full of unpaid-for random goods anyway. I usher all three children into the disabled loo because there is more space for us. Isaac is enjoying his freedom and tries to unlock the door several times. Molly sprays water from the cold tap all over the mirror. I help Bea with her school tights and tell her not to touch the sanitary bin as she lifts the lid and asks, 'What is this for?' When we are done, I know I can't face another minute of shopping, so we find a checkout. I have definitely forgotten several things we need, but I figure we have enough to keep us going and I can't wait to get out. I set the girls to passing each other packets, boxes and tins to load onto the checkout belt, being quick to grab things like eggs and jars of jam that could break easily.

I pop Bea onto the end of the till with a carrier bag and she puts a couple of packs of dried fruit into it and then starts to become swamped in tins, packets and jars as they beep past me faster than I can deal with them. I look up and smile at the cashier and say, 'Wow, it's hot in here this afternoon', and she looks up, smiling at my 'helpers', but doesn't slow down her robotic style of scanning goods. Three people wait in the queue behind me.

'That's £147.53, please', she says, adding, 'Have you got your points card?' I scrabble around in my purse full of store cards, receipts, loose change and notes.

'No, I can't find it,' I snap. I hate store cards. I hate that every shop has a different card, and that they think I have the time or inclination to fetch them every time, or that I'm interested in saving vouchers they send me in the post and then coming in specially to redeem 20p off a bottle of fabric conditioner. Just make the shopping a fair price and stop playing games. She taps her fingers on the till, waiting for me to gather everything up into the trolley, saying I can save my receipt and take it to customer services next time. I nod as if that's going to happen. I push my credit card into the machine and press my four numbers on the keypad and she passes me the mile-long receipt, which I scrunch into my pocket. I think: 'How did I just spend that much?' And: 'What on earth have I bought?'

I lift Bea down from the checkout and Molly nips her on the arm. Bea wails with pain. I feel like everyone around us is looking at me. I grit my teeth to get away from this place, and I give Molly the 'I'll deal with you later' look. The girls have run off ahead to jump in the Peppa Pig ride in the foyer. 'Not today,' I say firmly, looking again at Molly as they stand waiting their turn for a ride. 'I can't believe you just did that to your sister.' Then there are more tears from Bea because I said no; they watch a child laughing as he bumps up and down in Daddy Pig's shiny red car and Bea starts to howl as if someone has cut off her arm.

I struggle to push the trolley out of the shop. Our car now seems about five miles away and is up a steep slope. I pull their hoods up as we walk, telling them to hold on to the trolley as we cross through the traffic in the car park. It is raining heavily now. From the top corner of the car park I can see a man carrying a sandwich to his white van parked in the parent-and-child spaces and have to stop myself from going over and savaging him. Back at our car, the shopping sits in the trolley outside getting soaked. I open the boot only to see that it is still full of recycling I should have done before we got to the shop. I reorganise all the stuff, telling the kids it is like the Julia Donaldson story *A Squash and a Squeeze*. I fill the footwells with washing powder, bags of flour, sugar and cartons of juice. I check the children have all got their seatbelts on. Molly says she is hungry. I get out of the car again and rummage for something in the boot. I pass them snacks and we sit in the steamy car, not able to open the windows as it is too wet outside, and munch on chocolate-chip brioche rolls. I poke little straws into cold Happy Monkey smoothie pouches for them, begging Isaac not to squeeze his all over himself. I back the car out and drive home. I still have no idea what we are going to have for supper.

## Broth

I carefully carry a tray of steaming bowls of ham hock broth and set it down for the farmers. Plates of sausage and apple lattice pie and chunks of bread are already dotted along the long trestle tables. 'Butter!' I remember, and dash back to the kitchen to get the little saucers of slightly softened butter squares. There are glass tumblers set out and bottles of water on the tables to refill. I send Molly back and forth to the kitchen with empty bottles and bowls and ask her to fetch more plates of pie. The first few farmers start to eat. Fat snowflakes are falling fast out of the sky, blown at the glass doors at the end of the shed. It will be deep snow soon. Men and women are still arriving in their farm trucks; a couple of our neighbours have come on their tractors.

I am feeding seventy people. It started off as a small event, just twenty or thirty, we thought, but then a few more were invited, and then a few more. 'No more!' I said on repeat. But farmers like parties, so the list got longer. I have cooked and baked and cleaned and set up for two weeks. I made broth because it is cheap to make and is a hearty meal in a bowl. It is also easy to heat up and serve on the day. Our farming friends have come along to a charity stock-judging event we are holding in the barn. It is a Saturday lunchtime in December, and I am amazed that they are still arriving despite the treacherous driving conditions.

### SEVENTY-FARMER HAM HOCK BROTH

Broth makes you feel warm and nourished. Cooking it for someone else tells them you care about them. It is also very old-fashioned healthy food. It is cheap to make because it transforms basic everyday ingredients – bones and root vegetables – into a highly nutritious meal. The ham hock is a low-value bit of the pig, basically a knuckle joint. From one hock and a few vegetables you can make enough broth to feed a roomful of people – simply scale it up for even more, like I did when I found myself hosting seventy farmers! It is so simple to make that anyone can do it – it is little more than chopping vegetables and slowly boiling them with the hock.

When I'm planning a broth, I focus on making sure the simple ingredients are the best I can get hold of. I try to choose outdoor-raised free-range pork (ideally from our own pigs) and good locally grown vegetables (or, better still, ones from our garden). I make broth two or three times a year: in early autumn, around Christmas, and always before our lambing season starts in April. I freeze portions in saved tubs and pots to bring out on bitterly cold days to defrost on the Aga while we are busy amongst the sheep and lambs. We can quickly heat it up and enjoy a nourishing meal in a bowl for a hearty lunch or supper. It is such a forgiving recipe that, if I don't have exactly the ingredients listed overleaf, I can substitute the vegetables for other kinds. Parsnips go well in it, or diced celeriac. I also make it with a mutton bone sometimes.

It's a farmhouse recipe that has passed down the generations and everyone can get involved in making it. As it cooks, the house fills with the glorious smell of simmering ham and vegetables. Broth is almost like medicine. It seems to stave off coughs and colds and it builds up our immune systems for the hard days of winter work on the farm.

The recipe below is for one pan of broth and uses one ham hock and four pints of water. I usually make two large pans and simmer two hocks in each, as once I have committed to making broth it is just as easy to double or quadruple the recipe and freeze portions for the weeks to come. The effort involved will be rewarded weeks later when you enjoy steaming bowls of goodness, knowing you've made it from scratch.

# HAM HOCK BROTH

Prep 50 minutes

Cook at least 5 hours

Makes 1 large pan (serves 6–8)

## Ingredients

1 ham hock (outdoor-raised pork, if you can), approx. 1kg/2.2lb

200g/7oz dried soup mix or dried pulses of your choice (barley, split
   peas, lentils)

4 carrots

2 onions

½ turnip

3 celery stalks

1 leek

good glug of olive oil, plus extra to serve

organic or home-made chicken/vegetable stock on standby, to top up
   if necessary

kale, cavolo nero or spinach, to serve (optional)

parsley, to serve

squeeze of lemon, to serve (optional)

## Method

1. In a large pan or in a slow cooker, cover the ham hock in cold water –
   about 2¼ litres/4 pints should do it.

2. Bring the water to the boil and simmer gently for 2 hours, or 4 hours
   in a slow cooker on high, until it looks like the ham is starting to fall off
   the bone.

3. Remove the ham hock and leave to cool. Skim off any white foam or
   fat from the top of the water and strain the liquid through a sieve into a
   large bowl. [This can be made up to 3 days ahead and the ham hock
   stored in the fridge, covered, along with its stock.]

4. Using the large pan or slow cooker again (no need to wash), pour the
   ham stock back into it and add the dried soup mix or pulses of your

choosing. Cook for a further hour, gently simmering, or for another 4 hours in the slow cooker.

5. While all of that is cooking away, finely chop the carrots, onions, turnip, celery and leek – as small as you dare. I like to do all of this by hand and if I am making 4 times the quantity it can take a while, but you can enjoy that time by listening to a good audiobook or podcast. Get the kids to help or – even better – ask a friend over for coffee and put them to work with the promise of a pot of the finished broth as a thank-you. You can use a food processor, but the veg usually doesn't end up as tasty or attractive in the finished soup.

6. Using a good glug of olive oil in the bottom of another heavy-bottomed pan, sauté the vegetables until they are soft – be patient, this takes at least 20 minutes. Stir gently and season generously with salt and pepper.

7. Add the softened vegetables to the pan of ham stock and softened pulses. Set this to simmer gently and prepare the ham.

8. Using your hands, peel the chunks of meat away from the bone and discard the fat. Chop or tear the pieces into a bowl before adding to the stock pan, making sure there is enough stock to cover the ham and veg; add regular stock if needed.

9. Cook for 40–50 minutes, checking the pulses are soft and the broth is flavoursome. Add more liquid and seasoning if necessary. Broth improves over a couple of days – it will go thicker, so just thin it down to your taste with chicken stock.

10. When almost ready to serve, add greens and fresh herbs of your choice and simmer for 2–3 minutes to cook them through. Serve with a squeeze of lemon and a drizzle of olive oil if you like.

11. If not serving immediately, store, covered, in the fridge for 3–4 days or freeze in pots.

The valley outside the windows is transformed from the dirty greens and browns of winter into clean white fells and fields. I woke early to set up tables and chairs and wrap cutlery in Christmas napkins.

I have made three apple crumbles for pudding; we serve it school-dinner-style from giant trays in the kitchen, offering jugs full of custard and cream to go with it. My mother-in-law is firmly planted by the sink, washing everything that comes her way. I know she doesn't want to be amongst too many people. She'd rather keep busy in here than make small talk, because today is being held in memory of her husband, my father-in-law, Tom. He died from cancer ten months ago. These farmers are his friends. Each year they've taken it in turns to host the others, spending time on the farms, looking at each other's cattle and sheep and enjoying a meal together in the farmhouse or at the local pub. We want everyone to feel welcome and that means feeding them well.

My mum scoops the steaming broth into the bowls I have hired with a ladle from the huge pan simmering on the cooker. A drinks bar has been set up in the stable; men and boys carry bottles of lager or cans of cider around with them as they chat to each other. After they have eaten, they stand by the rails of the sheep pens and start to fill out their scorecards with little pencils. I have set up a tea and coffee table next to the dog kennels – two urns are boiling, ready to fill paper cups with teabags or instant coffee. Tubs and tins of home-made cake and biscuits start to pile up on the table as more people arrive. The other farm women have baked to help me out. Their shortbread and tea breads look wonderful. Lunch is an ongoing affair as we can't seat seventy people all at once. 'Take a seat through there and we'll fetch you some food through,' I say as ever more people arrive.

## BISCUITS

### SIMPLE SHORTBREAD BISCUITS

Prep 20 minutes

Cook 15 minutes

Makes about 20 biscuits

## Ingredients

170g/6oz unsalted butter, softened

85g/3oz caster sugar

225g/8oz plain flour, plus extra to dust

30g/1oz self-raising flour

## Method

1. Heat the oven to 150°C/130°C/gas 1 and place one or two baking sheets in the oven. Beat the butter and sugar until smooth.
2. Sift the flour in and beat to form a dough, but try not to overwork it.
3. On a lightly floured surface, roll out the dough and cut into 4mm-deep rounds. You can also do this by rolling the dough into a sausage and slicing off 4mm rounds. Any that you don't use can be wrapped, labelled and frozen.
4. Bake on the preheated baking tray(s) for 10–15 minutes. Don't overbake. They should still be light in colour but not golden.

## RAISIN AND OAT COOKIES

**Prep 10 minutes**

**Cook 12 minutes per batch**

Makes 24

## Ingredients

250g/9oz unsalted butter, softened

160g/6oz caster sugar

160g/6oz dark soft brown sugar

2 eggs

¼ tsp vanilla extract

380g/13oz plain flour

1 tsp bicarbonate of soda

a pinch of salt

½ tsp ground cinnamon

110g/4oz rolled oats

220g/8oz raisins

## Method

1. Heat the oven to 170°C/fan 150°C/gas 3. Put the butter and sugars in a mixer and whisk until light and fluffy.
2. Add the eggs and vanilla and beat well, scraping the sides of the mixer halfway through.
3. Add the flour, bicarbonate of soda, salt, cinnamon and oats and mix again.
4. Stir through the raisins.
5. Scoop out small balls (roughly 30g/1oz) of cookie dough onto a lined baking tray and press gently with a fork.
6. Bake for 10–12 minutes until golden brown – you may need to do 2 or 3 trays or bake them in batches. (Rolled cookie dough can be frozen on a tray and baked from frozen for an extra 2 minutes.)
7. Leave to cool on a cooling rack.

Our barn is set up like a sheep show, with two breeds of sheep on display. Our neighbour has brought some classy Swaledales. We have supplied some fine Herdwicks. The pens were all cleaned out and bedded with fresh wood shavings, with clean water buckets and tidy hay racks. It all looks strangely neat and tidy. We have three classes of sheep, of different ages. There are four sheep in each class penned together – ewes, shearlings (two-year-olds) and gimmers (last year's lambs). All of them are breeding females that will spend their whole lives on their farms, and they're rarely seen, so the shepherds love having a nosy. They have been prejudged by a master judge for each breed – two of the most respected older shepherds – and everyone has an individual scorecard to list the sheep from first (best) to fourth, based on how close to their ideal for the breed they think they are. After everyone has done their judging, the cards are gathered in and marked. The master judge gives his scores aloud and his reasons for his decisions, and then the person with the nearest score to the judge's decision is the winner. A local butcher has sponsored the event and has given us a huge hamper for the first prize.

I clap my hands loudly a couple of times. They all look towards me standing near the kitchen with my apron on. 'Right, I need your attention . . . everyone!' I thank them for coming, despite the awful weather. I say, 'It is nice to see so many of Tom's friends here.' I explain that we want to raise money for the hospice that looked after him so well, that we will be holding an auction later and that I want them to help themselves to tea and coffee and cake throughout the afternoon. But I also tell them I need their help with something else. I can see a few puzzled faces around the shed. In exchange for feeding them all day, I want them to put their handprint on a sheep. No, not a real sheep, I explain. I have a white moulded sheep upstairs in the loft. I know art projects aren't things most farmers would get involved with, but a few faces relax as they realise what I'm asking them to do is quite simple. 'Stick your hands in some paint and then on the sheep, and then it's over and you can have a drink.' There are going to be about a hundred of these sheep dotted all around the Lake District. Each will be different, individually painted or decorated by an artist. I tell them I'd like them all to be a part of my piece. 'I want you to handprint the body of the sheep because I want to call my finished piece *In Safe Hands*.'

'Here, dip your right hand into this tray.' A seventy-year-old farm-er I know well takes his smart tweed jacket off and hangs it on a chair, then rolls his shirtsleeve up and presses his hand into the bright paint. 'Ooh, that's cold,' he says, smiling at me. 'Move it around, make sure you've got enough on,' I say before he cautiously lifts his hand from the tray and holds out his purple-smeared palm to me like a child. I show him roughly where to press it onto the life-sized model of the Herdwick sheep. 'Find a space on the body of her,' I say, 'and press it down hard but lift it off gently.' He peels his fingers off carefully, leaving his mark. It shows the cracks of his skin, and an impression of his bent thumb. 'Well done!' He looks around the sheep, looking at the other marks that have been made.

Soon the body of the sheep is covered in multicoloured hand-prints. I also ask the farmers to handprint a sheet of thick paper, which helps take the excess paint off before they wash their hands, and they sign their names on the paper as a record of who came and took part. These hands are familiar working hands, rough, weather-beaten, thick and strong. Some are missing a finger or a nail from some farm-related accident. Some are bent in strange ways, or curled as if used to holding a shovel. These hands tell stories of the work the men have done, the work that keeps these old northern fell farms going. These hands have built drystone walls, birthed lambs, lifted fallen tree branches, drowned kittens, hauled bales onto trailers and cast hay across the fields. They are hands that have fixed icy water pipes in the mud and muck, and milked cows and stroked sheepdogs by the fire.

'There's a bowl over there of hot soapy water to clean yourself up,' I repeat, as each one comes along and joins in the fun. 'Thank you!' I have a queue forming of men and women who are ready to press their hands into the trays of red, yellow, blue, purple, pink, orange and green paint. All the kids are laughing at the farmers' reactions to my unusual request. One of the older farmers, who thinks it is all great fun, says, 'It's like being back at nursery school!' There is more laughter around me: some of the younger lads are pushing each other, threatening to put their hands onto each other's faces or slap one of the old guys on his back with a bright-pink hand, but I know them and when it comes their turn they do it thoughtfully.

The finished painted ewe looks brilliant. I love her. The colourful paint is similar to the smit paint that we use as shepherds to identify our individual flocks. Seventy shepherds have placed their hands on her, and she is a symbol of the care that these flocks in this ancient landscape receive. I leave her to dry in the loft and go back to the kitchen to help with the washing-up.

We auction off chances to take ewes to prizewinning tups and pass round a bucket that comes back full of twenty-pound notes.

We sell raffle tickets for various prizes of bags of feed and bottles of wine. Deep snow has now buried everything outside the barn. The kids are running around playing in it. A few older farmers from further afield head home, aware that they may have to pass over moors and mountain passes in the blizzard. The rest sit around and chat, and laugh, and get more and more drunk. Later I heat up some chicken curry I made for supper and serve it in bowls with some naan breads I bought. A bottle of whisky is opened. The fire is roaring, and I leave James with a group gathered around the sofas at one end of the shed, singing old songs and telling stories.


## Chocolate Cake

It is a sunny day in July. The scent of meadowsweet is heavy in the air as we park our car and walk down the long path through a leafy wood with the lake in the distance. A perfect day for such a sad occasion. Hushed voices all around us and respectful smiles and nods to older folk we know who walk slowly with sticks, as we make our way to the little sandstone church that has been here since the beginning of time. I lean into James's side, my heels sinking slowly into the mossy ground. He is wearing a black suit and his best shoes. I have chosen to wear a bright coral-coloured cardigan over my smart navy dress, because today is about celebrating a life as much as it is about mourning the end of one. We are at the funeral of my friend's mum. She died from cancer.

The coffin arrives and her sons-in-law carry it high on their shoulders. Her four daughters follow, quietly sniffing into concealed tissues clutched tightly in their fists. My friend is amongst the huddle, her arms linked through her two daughters' arms; they all wear bright flowery-patterned dresses. Her girls are the same age as mine. They were at our house playing on the day their grandma died. I smile at her, holding my lips together tightly, and nod my head once as if to say: 'You can do this.'

The church organ plays loudly, filling the air with notes, and I hold our folded Order of Service tightly, waiting to sing familiar songs. The church is so full already that we have to stand outside listening to the speaker by the door. There are a hundred people or more standing around us on the grass.

I strain my ears to listen carefully to the vicar, whose soft voice tells us of the life of this woman. Where she was born, where she went to school and then about her getting married and moving to the farm to bring up her family. She talks about her love of baking and how she won prizes at the local shows for her sponge cakes and scones. How she tended her kitchen garden and always put out wonderful spreads of food. She tells us that, despite not coming from a farming background, my friend's mother soon found her feet and made a great job of rearing the lambs and calves. She kept a flock of hens and was always outside helping her husband on the farm. We hear little stories of her taking the bus to watch the football with her grandchildren in the city, because she never learned to drive.

A warm summer breeze lifts through the trees, rustling the green leaves. A stream runs beside the church and trickles with a quiet flow to the lake in the distance. We are standing near the old front door, which is wide open. I look up to see the little bell mounted in the spire. People have gathered here at this church since before the Normans came in 1066. It has an ancient stillness around it. Sheep graze in the meadow over the wall as they have always done. I step to the side slightly to take the weight off my toes in these heels.

The vicar describes a woman whose whole life has been about her family and friends and how she cared deeply for them and the land and the animals they farmed. Her love and work mattered to so many people. The church and churchyard are overflowing with those who have benefitted from this humble women's kindness and care. I can't imagine a life spent in a better way, and for us to say goodbye to her in such a gorgeous place is sad yet uplifting. The sound of her granddaughter's familiar voice on the speaker breaks my trance. All eight of her grandchildren say a few words about what they liked best about her, and how good her chocolate cakes were, and they share their favourite memories of their grandma.

## EASY CHOCOLATE CAKE

Prep 25 minutes

Cook 40 minutes

Serves 8

Ingredients

For the cake:

175g/6oz softened butter

175g/6oz soft brown sugar

3 eggs

150g/5oz self-raising flour

1 tsp baking powder

30g/1oz cocoa powder

50g/2oz dark chocolate

1 tsp vanilla extract

2 tbsp milk

For the icing:

225g/8oz softened butter

225g/8oz icing sugar

225g/8oz dark chocolate, melted

Method

1. Heat the oven to 190°C/170°C/gas 5. Grease and line 2 x 20cm sandwich tins.
2. Whisk the butter and sugar together in a stand mixer or by hand until light and fluffy and the sugar has dissolved into the butter.
3. Add in the eggs, flour, baking powder and cocoa powder and whisk again until combined.

4. Melt the chocolate in a heatproof bowl over a pan of barely simmering water.
5. Stir the vanilla, milk and melted chocolate through the cake mixture until fully combined. Split the cake mixture between the tins and smooth down gently with a spatula.
6. Bake in the centre of the oven for 20–30 minutes until risen and a skewer comes out clean.
7. Make the icing. Place all the ingredients in a food processor and whizz until blended and smooth. A hand whisk or stand mixer works well too.
8. Allow the cake to cool completely before icing, then use half the icing to fill the cake and half to top it.

I smile at James and a feeling comes over me that I have never experienced before . . . It is as if the vicar is talking about my life and I have stepped outside myself to imagine that this is my funeral.

There are many ways to live, many ways to be a woman. I know lots of women don't want what I want. Some would say mine is a small life. But this is how I want to live my life. It is my choice. I chose the good and the bad of it. I look for a tissue in my bag and dab my tears away as they roll down my cheeks. James squeezes me close to him.

The sun is now at its height over us and I feel light-headed from not having eaten before we set off. The coffin is carried out and away to the car to go to the crematorium later this afternoon. This lady wanted to have her ashes scattered by the lake near where they farmed, which overlooks this church. We take our time to walk up the dark leafy woodland path – I am glad of the shade – to the country house where the funeral tea is being served. While James chats to friends around us, I get a plate of sandwiches and cakes and find a seat in the shade to eat them. Lost in my own thoughts, sipping hot tea, I watch the bees humming all over the lilac bush.

I am thirty-eight years old. I have been wondering about having a fourth child for a few months, always swaying between 'I have three healthy children, what if something goes wrong?' and imagining living with regret, wondering what that little person would have been like and how he or she would have fitted into the family. A head vs. heart debate. I've looked to others to help me decide, but they can't. As James and I walk back across the fields, looking back to the lake and the distant fells, I no longer care what anyone else thinks. I know that I want to become a mum one more time. Practically and financially it's a crazy idea. Surely it's time to enjoy the life we've created. But I'm also not sure that I've noticed all that time going by at full speed. If we are fortunate enough to conceive, a baby will force me to slow down, to sit and feed, rock them to sleep and notice all of those incredible stages of walking, talking and learning to read, all over again. There are six chairs around our dining table. I want to fill the empty one.

Tom is born in the September just before I turn forty.

### Creamy Pasta

The sky looks like a heavy blanket waiting to smother us. The snowflakes fall gently, like we are in a snow globe, but this will soon be a disaster on the farm. It is definitely no idyllic winter-wonderland scene where rosy-cheeked children pull wooden sledges along whilst wearing pretty coloured hats and scarves.

Two weeks ago, when the forecast predicted snow, I rushed out to stock up on essentials like milk and bread, but it turned out not to be that bad. The roads stayed clear, the shops were all open and the schools didn't close. So when more snow was predicted I didn't panic. I told myself it would be OK, and I could pop to the farm shop nearby for extra supplies if needed. But right now I know I am not going anywhere in this, definitely not with a baby. The snow is coming in heavy and fast. We will run out of the basics very soon. The news on the radio warns of a storm; they are calling it The Beast from the East.

We watch from the window and I can feel the anxiety rising in us both. James pulls his waterproofs, wellies and thick coat on, and I pass him his gloves, hat and neck warmer, gathering them up from around the fireguard where they have been drying out. I hold Tom up under his arms, his legs kicking in his Babygro, to wave his daddy off at the window. He is five months old. The quad bike chugs and cuts out. It takes James three goes to start it – the engine is stone-cold. With the bike running, he hooks the trailer onto the towbar and calls to the four dogs. They race each other, falling over in the snow. Bea is following – she went ahead to let them out of their kennels. She perches on the side of the bike with her dad and the dogs chase after them down the lane as they set off to shepherd and feed the flock.

I sit with Tom on my knees by the window and bounce him up and down, singing 'Wind the Bobbin Up' as we watch the snow fall. I think through what I need to make ready for James and Bea coming back in. They will have frozen hands and will be shivering despite their many layers. I fasten Tom into his bouncy chair and hang up a fresh toy for him to try and reach, then light the fire. The log basket is nearly empty, so I ask Molly to keep an eye on Tom while I take it to the garage to refill. On my way I take a bundle of dirty baby clothes and towels I've left on the stairs to the washing machine and the new pack of nappies to the bathroom, unload the machine, hang the cold wet clothes on the drying rack and put another load on. In the garage I fill the log basket and rummage in the freezer to find pastry, chicken, a packet of beef burgers and some bacon. I load it all up on top of the logs and carry everything to the kitchen, leaving the food on the worktop to defrost for our next few meals.

On my way downstairs, the lights go off. The electric must have tripped. I go back to the garage to check the fuse box, but all the switches are on. There is just no power.

'Power cut!' Molly shouts from downstairs.

'The generator hasn't kicked in,' I reply. 'You'll need to get the ponies in before it gets worse.' She tells me that they have rugs on and will be fine. I tell her insistently that this could be the worst snowstorm this farm has ever seen. 'Go and get them in,' I plead, 'and if you see your dad, ask him to start the generator.' I hope we haven't run out of diesel or this really will be a disaster.

Living on the side of the fell with no mains electric means we often have power cuts to sort out. The solar panels are now thickly covered in snow so either the generator hasn't started automatically or there is another issue with the battery storage system. I know this is not good news, but it is only 10 a.m. We have got time to sort this before it gets dark.

As I move around the house I can feel draughts blowing in from every nook and crevice. The windows and doors rattle with the wind. I tuck a blanket around Tom, who is busily chewing on his teething ring. He is watching the flames of the fire, which hasn't got any real heat in it yet. I load up the log burner and ask Isaac to move out the way to let me change Tom; he has a whole blanket spread out on the floor with Lego and is busy building a fortress.

James comes to the back door looking like a yeti. He shouts to Molly, 'I need help *now*.' He is cross as he shouts to her: 'The ponies need to be in their stable. Bea has gone to feed and water the pups in the shed.'

The snow is blowing into the kitchen. I shout, 'Shut the door – we need to keep the heat in.' I add gently that the power has gone off and ask him to check and start the generator.

Molly layers up, pulling my old green waterproof trousers over her clothes and fastening her coat up fully. She rushes to follow James up the driveway, struggling to run in her layers. The snow is piling up around the car and has made rippled drift already. I click the gas burner on and light it with a match. I need to find the box of emergency oil lamps and candles just in case we can't fix this. I find the box I need on a shelf in the pantry and go upstairs to pull out extra blankets, ready for the night ahead. I can hear Tom fussing so go to settle him. He has had long enough in his chair, so I change his nappy on the floor in front of the now roaring fire. I let him kick around while I go back and forth to the gas hob and pan-fry cubes of chicken and leeks – it will have to be a creamy pasta sauce now that the oven is off. I boil some water in a pan, make a cup of tea and use the rest to sterilise two bottles for Tom for later.

## KITCHEN TABLE CHICKEN PIE

This recipe will make one family-sized pie in a small roasting tin or ovenproof dish. Alternatively you can cook some pasta and add to the filling for a creamy pasta dish.

**Prep 20 minutes**
**Cook 50 minutes**

Serves 6–8

### Ingredients

500g/1lb free-range chicken – I like to use half breast and half boned and skinned thighs, cut into chunks. You can also use cooked meat stripped off a roast chicken if you have any leftovers – just add it once you've made the white sauce.
50g/2oz diced bacon or pancetta
a few mushrooms, sliced or quartered (optional)
1 large leek or 2 small ones, cut in half lengthways then into 2cm slices
300g/10oz puff pastry (ready-rolled or rolled out from a block)
1 free-range egg, beaten

### For the white sauce:

570ml/1 pint whole milk
50g/2oz butter
50g/2oz plain flour
1 organic chicken stock cube

## Method

1. Heat your oven to 180°C/fan 160°C/gas 4.

2. If using raw chicken, fry the meat in batches over a medium heat until golden and cooked through. Remove the chicken from the pan and set to one side.

3. Fry the pancetta or bacon in the chicken pan and add to the chicken. If using mushrooms, cook them for 2–3 minutes in the bacon fat before adding them to the chicken too.

4. Fry the leek or pop in a microwaveable dish with a small knob of butter and a splash of water and cook on high for about 4 minutes, until soft.

5. Make the sauce by putting the milk, butter and flour in a clean pan and heating through gently, stirring all the time with a whisk until it has thickened. Crumble in the stock cube or, if your sauce is looking thick, add a little boiling water too along with any juices from the chicken or leeks. Combine the creamy sauce with the chicken, bacon, mushrooms and leeks. If you're using cooked chicken, add it at this point. Stir gently to combine it all and spoon the mixture into an ovenproof dish or roasting pan.

6. Roll out the pastry on a lightly floured surface to 5mm thick. Wet the sides of the dish or pan with just your fingertips.

7. Carefully lay the pastry on top of the filling. Trim excess pastry with a serrated knife and crimp the edges with your fingers. Use any leftover cuttings of pastry to make decorations, if you like. Cut 3–4 air holes in the top to let any steam escape and brush with beaten egg. Bake for 20–30 minutes until crisp and golden. I usually serve my chicken pie with roasted, mashed or new potatoes and steamed or boiled carrots.

I stand at the door watching the fat white flakes engulf everything. My phone rings. It's Mum.

'Are you all OK?' she says. 'It's a bad forecast. Have you got any snow yet?'

I say, 'Yes, it's really heavy here. James is out in it with the girls, but we're fine.' I don't tell her the power is off; she'll just tell Dad and he'll mutter on again about us not spending the money to put mains electric in when we moved here (despite it costing an absolute fortune that we didn't have).

She tells me that my brother-in-law is stuck in traffic down the country in his wagon, and my sister is on her own tonight again. I think I'm supposed to feel bad for her with this news, but I can only think of what I have to do before it goes dark to keep us all safe and fed with no electric. I am not sleeping well either, as Tom is up every night at least twice and it makes me sharp with my reply: 'What does she expect? It's his job.' Mum chatters on about her dog instead and says, 'Bye for now – ring me if you need anything.' But I have to dig in and get on with this. There is nothing anyone else can do.

I feel bad after the call. I know she was only trying to help and that my sister is stressed out too. I'll send her a message later. Sometimes my only way of coping is to buckle down. I know I can come across as a bit mean.

I remember that I made myself some tea. I go to sip it and it's cold, and as I put it in the microwave I feel instantly stupid as there is no electric. I turn to the hob and light a match to the gas to make a fresh one. I fill a flask at the same time, ready for James.

When he gets back forty minutes later, he tells me the generator won't start and the batteries are completely flat. He wants me to ring an engineer to come out. He has to go back to check the cows. I tell him to have a drink to warm up, but he doesn't have time. I call the engineers but they can't get a van here today; they're all out on jobs already. 'Tomorrow, maybe,' the woman on the phone says. I stress to her that we are on a farm with a five-month-old baby, and it is blowing a blizzard and we have no power at all. Just as she starts to speak again, the phone cuts off. We don't have a landline at the farm and rely on our two mobile phones, so I take a deep breath, knowing now that we have no power, no phone signal and no internet. At least I told her the facts quickly. They will send someone

when they can. The fire is doing a good job of keeping the living room and kitchen warm, but the rest of the house is going cold fast.

I wait for James, nursing Tom in my rocking chair. Isaac is still busy playing Lego. I watch the window and listen out for the sound of the quad bike, but it doesn't come. Tom falls asleep on me as the snow falls.

I whisper to Isaac to pack a few toys and get his pyjamas and toothbrush. I know that we can't stay in the house with no electric – it will be too cold in the bedrooms. We need to go up to the sheep shed/office to sleep tonight. It is well insulated and once that fire is on it is as warm as toast, and we can all sleep in one room. Isaac finds this idea hugely exciting. I think through what I will need. Last time we ran out of heating oil it was a bitterly cold night even with several blankets. Now we have a baby and there is a snowstorm out there. I set Tom down in his bouncy chair and gather up the chicken and leek sauce I have made and a bag of pasta, thinking I can load it all up in the car and drive Tom up to the top of the lane, but the snow has now buried the car in a thick drift. We will have to walk it.

James appears through the blizzard at the door again. His quad bike has cut out and really won't start this time. He has some sheep stuck on the other side of the farm that can't get to the hay he has put out. They have now huddled behind a wall and won't move, even with the dogs. I say he'll have to leave them. I tell him my idea, that we need to up sticks and move into the sheep shed. We can run the little petrol generator he uses at shearing time, with an extension into the office to at least charge a phone and power a lamp or two. There is a heap of firewood up there and it is well insulated so we will be warmer. Molly and Bea come back in shivering, but they think this is a great plan. They strip off their snow-caked boots and coats and rush upstairs to gather what they need.

As I look past the garden to the fells, everything is white, like we've stumbled into Narnia. The snow keeps falling, more gently now, but the fat flakes soon melt on us and make our coats wet. I strap Tom into the baby carrier on my front. He is kicking and wriggling with all the excitement. We are shouting as if we can beat the sound of the wind.

'OK,' James says, 'I'll get the fire lit up there and come back for you and the boys.' The girls have packed a bag each and tell me they are going up, pulling their stuff on sledges, and will see us soon.

Isaac helps me as I put general stuff we need – matches, cereal, milk, a loaf of bread and some butter – into a basket. He passes me the frozen drinker from the guinea pig hutch that stands beside the kitchen door and I defrost and refill it. I throw an old rug and a blanket over the hutch for extra insulation, pinning it down with a couple of stones I pick from the wall, brushing the snow off. They'll be fine, I tell Isaac, who wants to take the two hairy creatures with us. 'Look, they're snuggled into the hay,' I say. 'And they have food and water, but go and get an extra carrot for them if you want.' I wade through snow on the patio to the hen hut to see if they too have enough food and water. Tom is wrapped up in my big coat on my front, peeping out with his nose and cheeks quickly turning pink.

We have everything by the door when James comes back to help me. I pass him two bin bags of clothes and towels and cover up the basket that I have put food in. I put the basket in the three-wheeler buggy and struggle to push it all up the lane with Tom trying to kick his way out of my coat, but I dig in, and push harder, because once we are there it will all be alright. We hike up the hundred metres to the sheep shed with our random bags of stuff. The pups yap at us as we pass them in the barn; they are four weeks old now and bounce up at the railings of their straw pen. Tom is oblivious to all of the drama. He watches Isaac and giggles when he tickles his feet. I set him down on the office floor with a few cushions around him for support and unpack the bin bags, pulling the two mattresses out from the back of the office to be nearer the fire. Isaac jumps up and down on one of them, claiming it as his bed.

James, Tom and I will sleep on the other. I pull sleeping bags out for the girls to use on the sofa bed and pile the cushions up for extra pillows. It is a huge family sleepover, I tell Isaac, smiling, trying to hide my worry. The wind is howling now.

I think back to the floods only three months ago in December, when Storm Desmond hit the valley. I had to pack three-month-old Tom up in the car with the other kids and drive down the road through deep puddles, then leave them all in the car on high ground for an hour while I walked two hundred ewes from the other side of the river back to higher ground with a shepherd friend. James was abroad working and on his way home, but the flood wouldn't wait for him and the sheep were in a dangerous spot, at risk of being washed away. 'I can manage this,' I thought. We are all safe and well together, we will be OK.

James is working in the shed. He tries to start the petrol generator we only use in the summer. It is slow to get going. He wrenches at the pull-cord, one, two, three, four, five times, getting more and more angry with the situation we are in, and then the pull-cord snaps. Now we really don't have any power for the whole night ahead. I'm cross with him for being so rough with it and take the extension cable that I found and roll it up again. I walk away from him, cursing under my breath, set out the candles and prepare for the coming night. Tom starts to cry, and I realise I haven't fed him in ages. I find his bottle of milk in the bag and warm it in the flask of hot water I fetched up.

It is getting dark by 4 p.m. and outside it is deathly quiet. Snow has muffled everything, and it is still falling, silent and deadly. It reminds me of my favourite *Blackberry Farm* story I often read to Isaac, when the animals watch giant snowflakes plopping down from the sky and the owl tells the rabbits how bad it'll be in the morning. I fear for the safety of the sheep and cows and hope they don't get buried in it.

We settle in for the night, trying to make it all seem like a wonderful adventure for the kids. James reads a story to Isaac, but snaps at me when I ask him if he has defrosted the water pipe for the sheep in the shed. He goes to fetch more firewood in. I know he has had a shit time outside and he is really worried about the livestock and frustrated because he doesn't know how to fix all of this. The storm is getting worse, and the sheep need checking and feeding in the morning. We don't even have a tractor of our own to put the bales out and have to rely on our neighbour to come and do it. I know how he hates that we have no money for one. We eat forkfuls of pasta mixed with the chicken and leeks that I have heated through and the girls get showered and dry in the bathroom that thankfully relies on gas for hot water. I feed Tom again and tuck him next to me on the mattress on the floor. James uses the candlelight to read by, saying he will stay up to feed the pups and check the sheep's water in the shed, and that I should try and sleep. Everything is worse when we are tired.

After a restless night, it is clear that we are completely snowed in. The snow has blown up in big drifts against the windows. The kids wrap up in their farm coats and waterproofs and race to get their plastic sledges and tear off down the hill. I tuck Tom into his snowsuit, mittens and hat and carry him to the front of the shed. The view is spectacular, the whole valley is completely white, but we have all sorts of dramas to deal with that stop me really enjoying it. I don't think the engineer will be able to get anywhere near us: the little road to the farm will be blocked with snow and our lane is impassable. We can't get the car out at all for any supplies. I am going to have to raid the freezer at the house again. I look at the quad bike that is now two-thirds covered in snow, and my heart sinks. James has trudged across the fields to ask our nearest neighbour a mile away for help. An hour later I can see a red tractor with a flashing light in the distance.

I watch James and Chris, our neighbour, unwrap and take three bales from the sheep pens to fill the hay racks in nearby fields. I check that Molly, Bea and Isaac are safe. They are chasing around with the dogs pulling their sledges up the steep hill. They shout for me to come and join them, but I have work to do. I mix the pups' milk with warm water from my flask, tip it into their shallow dishes, and after I stop them jumping in the dishes they lap it up. Tom squeals as they try to jump up at him, milky paws covering the baby carrier again. I scrape the dog pens out, fill them with fresh shavings and wheel the barrow to the end of the shed.

I pull my coat and boots off and busy myself in the office, with Tom still in the buggy watching me as I clean up, fold the bedding and load the log basket up again. There are puddles of melted snow in the porch to mop up and it will soon be lunchtime. I have brought bacon, eggs and butter back with me and get it all set out to cook. I boil another pan of water and fill the two flasks I have with coffee. James gets back in, exhausted from his hike up the lane in deep snow, and reports that the flock is safe but we still can't get

out. The road will stay blocked until a snowplough turns up. Chris has got his own farm work to do now but will come up and see if we need anything later.

I know that the engineer won't be able to come. I look around at the wet, cold, muddy sheep shed that yesterday looked like a sanctuary from the cold house and start to sweep up, getting cross that everything is in such a mess. Petrol cans, baler twine, old feed bags, shavings and hay strewn everywhere. James and I glare at each other until he puts his coat back on and goes out to check on the kids. I go to play with Tom by the fire and two minutes later they all pile back in frozen, caked in snow, with runny noses and rosy cheeks. I shout for them all to hang up their coats and waterproofs, not to leave them in a heap for me, and they raid the biscuit tin before I get a chance to ration them. My only way of coping right now is to try and create order out of the chaos, and anyone who makes that worse is in my firing line.

We have lunch and settle in for the afternoon. I suggest we play some board games. But after an argument over some issue Molly has with Monopoly, they all start moaning that they want to go back to the house and play and watch TV. I tell them to get a book, or do some colouring, but that idea doesn't impress them. They argue and wrestle until I ask them to go and feed the pups again, and I get up from the only sit-down I've had all day, as Tom is napping, to make the milk ready.

As it gets dark again, I heat two tins of soup and cook burgers in the frying pan I cooked the bacon in at lunchtime. I ask James to go back down to the house and look in the freezer for any meat, bread or rolls he can find. He brings two packets of crumpets and a roasting joint of lamb. I look at him with raised eyebrows and say, 'How on earth do you think I'm going to cook that with no oven?' We look at each other and start to laugh.

We spend another night huddled in our office. It isn't like a scene in *Little House on the Prairie*, all cosy and romantic – we have one candle and a tin of custard to share between the six of us. I have drunk so many cups of coffee that I'm wired and can't sleep again, and Tom is so uneasy teething that I can't rest even if I wanted to. Isaac starts crying that he has left his teddy back at the house. James pulls a book from the shelves and starts to read to him, Isaac's head on his dad's chest until he falls asleep.

It is the morning when Chris comes to check on us again, and I ask him to dig the car out with his front loader. James and I have decided it isn't safe for me to stay with the kids, with no power and no phone. I am going to pack up as much as I can and drive to Mum and Dad's.

He takes me to one side. 'Remember what it was like when we lived with them when Molly was born?'

'Yes, I know, it was challenging alright,' I say. 'Will you ask your mum if she'll have us?' We both know he can't leave the farm, especially with the pups, but he says he will be OK.

Tom waits strapped in his car seat with Bea entertaining him while Molly helps me pack up. Isaac cries again, saying he wants to stay with his dad. There is no electric here and there might not be for a few days. James charges his phone in the car while it is running, and phones his mum. We're lucky to have family so close by. I tell him I'll call the engineers to come about the generator as soon as I can. I follow the tractor out carefully in the tracks Chris has made and by then the snowplough has carved out a route for us to navigate. The road is five feet deep with snow in a sort of tunnel as we leave the valley, but when we get to the main road, three miles away, the dual carriageway has been gritted and traffic is travelling at a normal speed, as if our winter nightmare had never existed.


*Banana Peel*

I stare at the little digital clock in the car, which lights up as the engine starts. We have six minutes to get to school before the bell goes. I might just make it. I look back at the three of them in the back seat as I fasten my seatbelt. 'Don't spill crumbs, please,' I scold Isaac as he nibbles his toast. He drips honey down his fingers and looks at me like he is deeply sorry, but carries on. 'I've brought you a banana as well. I can't believe you didn't get any breakfast when I told you to.' His face is still, frozen between chewing and crying. I reach into the glovebox as I start to back the car out of the driveway and pass him the baby wipes to clean himself up. I soften my voice to stop him from crying.

Molly shouts, 'Stop! I need my PE kit!' I crunch the tyres to a halt on our gravel drive. She runs back to the house, and I say the same stuff I always say: 'I say every morning "Have you got everything?" before we leave the house, and every single time one of you forgets something. Bea, have you even brushed your teeth?' I look at her and she looks back at me guiltily. 'Well, they are your teeth, not mine,' I say, going back for another go: 'You're nine, that's old enough to know that you have to brush your teeth every morning.'

Molly bangs the car door shut and climbs over Bea's legs, and they argue. I snap, 'You could have moved across and let her in, Bea!'

'I don't want to sit in the middle,' she moans, and I can see her face sulking in the rear-view mirror as we bump our way down the potholed lane from our farmhouse to the road.

'We're late now!' I take a moment to breathe in deeply through my nose and then release it out my mouth, my chest heaving . . . and I plead, 'Please, please, can you all just stop arguing for a second and do what I ask you to do?'

I look at Tom. He is fastened into his baby seat next to me, the passenger airbag turned off. He is chewing on a teething ring and has dribbled all down his chin until his vest is soaked. 'Please pass me a muslin from the nappy bag by your feet, Molly.' I dry him and tuck it into his top as I turn the car into the village where the primary school is. There are barely any cars here now, so I pull right up to the space nearest the yellow lines. I wave to a friend who is driving off to work. And then I feign a smile to another mum beside her car. Her blonde hair is tied back neatly, she is wearing a black-and-pink slim-fitting top and she is busy tucking her fancy patterned leggings around her ankles before stretching, up and around with her arms, preparing to go running like she does most mornings.

I help Isaac out of the car with his bag and coat and the girls run off to the office, leaving him behind. 'WAIT!' I shout, but it's too late, they've gone.

'I don't want to go in on my own,' he says to me, nearly crying again, so I tell him to stand still by the car as I unclip Tom and manoeuvre him out of his car seat. I smile and nestle his rosy cheeks up to my face, kissing them and tucking a scarf around him that I grab from the footwell. The wind is chilly; I didn't put a coat on him before we left.

Isaac opens the creaking iron gate for us to file through, down to his classroom. 'I will call in at the office on the way back, Isaac,' I say, as I dab my finger in my mouth and rub at the syrup on the edge of his lips. 'I'll tell them to change the register, don't worry.' But I can see worry all over his little face. It is now 9.15 and I walk down to his end of the school shielding Tom and wishing we had been better organised. Isaac's teaching assistant is smiling and happy as she welcomes him into the classroom. I mutter my apologies and she asks if I am OK. I nod. 'Fine,' I say, smiling and shutting the door. I half-see myself in the glass as I peer in. I watch Isaac hang his coat up, and he tucks his packed lunch into the back of his locker as if to hide it. He walks into the room and sits down cross-legged to listen to his teacher.

I am left staring at my reflection. I have dark semicircles under my eyes. My skin looks pale and my hair is a mess. I am wearing my thick coat, torn from when I climbed a barbed-wire fence last week, an old pair of maternity jeans, the stretchy waistband pulled high up over my flabby belly and covered over with a bobbly grey jumper. My ankle boots are caked in soil and grit because we haven't got around to paving the front of the house yet and it has been raining.

I walk back to the car lugging Tom, and get him settled into his seat and ready to drive off. He will need changing as soon as I get back – his nappy was bulging as I strapped him into the seat. But first we need some bread, and I have two parcels to post, so we turn towards the next village. After a few miles I drive past the mum I smiled at earlier; she has run all this way, neat white headphones wired to her ears, her phone strapped to her arm. She is running confidently as I pass her. She looks in control of her life, strong, fit and ready to take on the world. I feel jealous and tearful. I angrily think, 'Who has time to go running after dropping their kids off at school? Every single fucking day?'

It is Tuesday and we have a farm inspection on Friday, and I feel anxious that I am not ready. I need to gather up all the paperwork and check that all our records are in order: the medicine book, flock health plan and dead list. I still need to update the livestock movement spreadsheet from last year. I know there are two people coming for a meeting with James this afternoon and I need to hurry back and do a day's housework before I pick the kids up again, and then work out what we are having for supper, all with a baby in tow. I glance back at the mum running. She didn't see me, focusing on her pace or something else. She looks determined but not fazed. She must be doing a loop through this village and the next before running back to her car, which must be about six miles. I have never run a mile in my life – I always joke out loud that my body would get a shock if I started running. I know I can move quickly if sheep escape and I need to get past and turn them. But I would never consider running as a pleasurable thing. I can't even manage to brush my hair some days, let alone go for a walk on my own.

When I get back from the errands, I change Tom's nappy and sit down to feed him. I check my phone. I scroll through a series of photos taken by a friend of the weekend away she has been on. She is in a restaurant, then a spa, then having drinks by a pool. I scroll on: another friend celebrating her birthday with a cake her kids have made her. Another friend wearing a medal in a photo, proudly standing with her bike after cycling the coast-to-coast route for charity. And another friend showing off her new haircut. I scroll, click 'Like' and scroll again.

I look at the weather outside. The window is scattered with tiny beads of rain. They start to form heavier droplets that wriggle down the panes. Tom sits on my knee and I rub his back gently as I tuck my breast back into my bra, adjusting the absorbent pad to be less scratchy against my skin, and clicking the bra-fastener with my finger and thumb. The ash tree in front of the house sways, its lowest branch swinging up and down in the wind as if it might crack and fall. James will be back in for lunch soon, and I haven't even cleared the breakfast up. I set Tom down in his bouncy chair. He watches me circle the room. It never stops being messy, this room where we live. Cereal boxes on the worktop. Isaac's Lego creations that I am forbidden to touch. Papers and unopened post piled up. Teddies on the floor. I open the dishwasher and it is full of clean dishes that need putting away. I think back to the mum I saw running two hours ago and suddenly want to be outside in the rain, running, running anywhere, running away, far from here, just me and the road and the rain.

Two weeks later, at the school Mother's Day service, the other mothers fuss around little Tom and take turns to cuddle him in the church while the children sing songs and pass out paper flowers to their mums and grandmas. I am wearing a clean jumper and smarter coat, jeans and boots, and the pram is shiny and new. To everyone else I look like I have it all together. I chat to the mum who goes running every day. As she smiles at Tom and tickles his toes, she tells me that I'm lucky to have four children and says, 'I don't know how you do it.' I smile, knowing she doesn't really want to hear my answer. She hands me a flyer about the marathon she's training for; I scan-read that she is raising money for a charity helping parents through infant loss. She is running the marathon later this year in memory of her baby who died when he was born. I choke up and feel sick to my stomach, ashamed of hating her that day in the rain.

There are all kinds of mums around me, and we are all carrying our own stuff on our shoulders. Unless we have walked a day, a month, a year in each other's shoes, how would we ever know what each one of us is dealing with?

EVENING

The hens come clucking and running to me at lightning speed when they see me with a pan in my hand. I scrape out the leftover rice from supper onto the grass with a spoon and they peck around my feet. Back inside, I wipe the table, put the last two glasses in the dishwasher, tip out the water jug and rinse it. I bend down and get a tablet from under the sink and put it in the dishwasher. My back is aching. The kids have made a good enough job of tidying the meal up, but there are still toys on the floor, boots and coats in a heap, papers on the island and more eggs than I can deal with on the worktop.

James is playing a game of dragons on the floor with the boys. They have built a tower with blocks and Tom wants his army of wooden hedgehogs to climb aboard his makeshift raft to escape the dragon castle. It's nearly seven thirty.

'Come on, Tom, it's bedtime. Do you want to go and hide upstairs, and I'll come and find you?'

'No! One more game,' he protests.

James lifts himself up off the floor slowly and says he'll do bedtime tonight as he can sense the weariness in my voice. Getting Tom upstairs is always a battle, negotiating the shift from him playing to his bedtime routine. He likes to hide behind his bedroom door and I have to act surprised when I find him. He has recently started getting himself undressed and wants to try to brush his own teeth. I am both happy and sad about these changes. The days of wrestling a small child into pyjamas are nearly over. He will soon be going to bed by himself and will no longer need tucking in with ten thousand kisses. But for now he still loves his story time and cuddles into my arm as I read to him most nights. He always wants just one more before I turn his light out. He knows I will give in because he is my baby. But tonight, I am glad of the free pass. 'Found you!' I say to Tom, chasing him to give him a tickle as he runs away from me. 'He's all yours,' I call down to James, and I pick up a discarded coffee mug from the sideboard on the landing. 'I'm going for a walk,' I tell him. 'It's a nice night, I won't be long', and as we pass on the stairs, Tom in his room, I kiss James on the cheek. 'Be good for your dad . . . I'll come and tuck you in later,' I call back up to Tom as I set the mug on the kitchen worktop.

I place my headphones over my ears and pull my hair out from behind my neck. I can feel a big knot in the back of it. I can't be bothered getting a brush and sorting it now. I'll have to get it cut soon it is getting too long. I click my audiobook on my phone to where I left off the other day, and tuck it into my pocket. Floss is around my feet, wagging her tail as I step into my wellies. As I walk out of the house, closing the door gently behind me, I look back at the chaos of our lives through the window. I can see that no one really notices I've gone.

Floss skips like a puppy down the field ahead of me with her tail up, as if she hasn't been out all day, but I know that's not true. She is an easy dog to have around in her retirement. Her one vice is that she barks and snaps at the delivery drivers when they turn up in their high-vis jackets, so I have to be careful. I like her company on my walks. For a couple of years after our beloved Bramble died, aged thirteen, I didn't have a dog to walk or with me in the house and I really missed that friendly presence. At that time Tom was a newborn. I couldn't even imagine taking on another pup to train and exercise alongside all my other jobs. I often look over to the beck in front of the house where Bramble is buried and think of her digging stones out from under the water and splashing around, me cursing her for bringing her wet muddy paws into the house, and her licking my hand as I stroked her afterwards, as if she was offering me an apology. She was a great friend.

I open the gate for Floss to slip through so she doesn't have to clamber over the cattle grid. We turn right out of our lane and along the quiet road towards the village. I press pause on my phone and look around. I have a good view of the front of the house and the beck that winds its way down the hill. Last week, when I had a school group visiting the farm, I saw some otter spraint on a rock in the middle of the beck and I pointed it out to them. 'There must be plenty of frogs and fish in the ponds and streams for the otters to eat,' I told them. The thorny scrub of trees and bushes that we planted nearly ten years ago is thickening now. The beehive is tucked in there too and I can just see the pale blue of its lid peeking out. James made the schoolkids laugh by doing a silly worker bee dance when he told them how far they go to forage for pollen and nectar. I told him that worker bees are all female and do all the hard graft. The male bees just buzz around the hive being noisy, mating with the queen and getting drunk on the sweet nectar. The kids laughed.

Two years ago we had the beck rewiggled – returned to its original course along the natural floodplain in order to increase biodiversity and reduce flooding downstream – and when I look at it now I can see a pair of oystercatchers pecking amongst the gravel beds; maybe they'll make a nest. It is still warm outside – I've tied my jumper around my waist and like the feeling of light breeze on my arms.

The grass is getting longer in the hay field now. We took the sheep and lambs off two weeks ago. When I explain to schoolkids and teachers that this hay meadow is a rare habitat, they look at me in disbelief. They only see a 'field' of grass until I explain that there are 110 species of different flowers and grasses growing here – we know this because we had a botanist survey it all. We have thirty acres of hay meadows here on our farm – close to 3 per cent of the total remaining upland hay meadows in the UK. Every field in Britain was like these once, and now fields are much poorer things, green factory floors. Most fields across the country only have one or two kinds of grass in them now. I then ask them to think, if they were a sheep or a cow, would they only like to eat one thing every day or would they prefer a variety? I tell them that, like us, the animals are healthier with a mixture to eat. And that food is also like medicine for them: they search out willows, herbs and different grasses and flowers when their bodies need them. The different species of plants support a whole host of insects and bird life. In this simple way we start to get children to learn about farming differently. That it's not 'farming or nature', it's 'farming *and* nature together'.

When we have schoolchildren on the farm it is important to me that they learn that their role (and their parents' roles) as consumers is crucial in all of this. Food doesn't begin and end with stuff from plastic packets in supermarkets, and fancy recipes on Instagram. It begins in the soil, in the plants and animals we grow to eat, and it ends in our bodies, affecting our health and well-being.

We spend time with the children digging up soil and having fun looking for worms, and explain that healthy soil should look like rich chocolate cake and should be alive, wriggling with worms. Not dry, dusty and dead. We talk about soil being the source of all life. We explain that we can't grow food without good soil, and we can't live well without good food. We talk about the role cows, pigs and sheep have to play in our landscape and why we farm the old-fashioned breeds of Belted Galloway cows and Herdwick sheep here on this hilly land unsuitable for crops. That these animals can withstand our Lake District long wet winters and turn this poorer land into incredible, healthy food. The children gasp with disgust as we lift up cow pats and talk about the importance of poo, then they get busy looking for the amazing shiny black dung beetles. We talk about the simple 'golden hoof' process of natural manure and why we need to move away from using synthetic fertilisers

and chemicals on our fields because they kill the soil and the life on the land.

On some school visits we explain the simplest version of our grazing pattern here. We now mob-graze our cattle and sheep in fields where we have let the grass grow much longer than we used to. Learning that the roots of plants in the soil mimic the height of the green plants in the fields was a game changer for me in understanding grazing ecology. If we graze grass short, it will have short roots. If we want healthy soil with lots of deep roots breaking it up, and pumping down carbon, we need to let the grasses and plants grow much longer. We give fields lots of time to rest between grazing them. Then, when the sheep and cattle do graze, they are mobbed up and in for a short time before moving on. This tramples the fields and adds that essential magic manure to the ground. The trampled matter they leave behind rots down, feeds the soil and covers it, like a blanket keeping it warm in winter. Our ground now holds much more water, helping with flooding too. It's a triple-win, for the land, the animals and us.

We also show schoolchildren the hedges we have planted, or get them involved in planting more, and we talk about why we need lots of hedgerows on farms. They are natural corridors for wildlife, giving food and shelter to birds, butterflies, hedgehogs, mice, bats and moths. Some children are amazed that we see foxes, squirrels, badgers, owls and deer here regularly. They often don't think these creatures have much to do with their lives, but we explain that they are all part of a giant ecosystem on earth and we are all responsible for looking after it.

It's easy to see how this disconnect and loss of knowledge has happened. I never understood all of this properly as a child; no one showed me the natural world the way James teaches our children. But it is our responsibility to learn about it and live more mindfully.

When I look across our farm now, I feel proud. The changes we have made are amazing, albeit with a crazy amount of hard work and paperwork, but we are enriching the land we live on. James makes me laugh with his ambitious projects for the future. He has a vision of our farm being half-covered in woodland, with great cattle and sheep grazing in the clearings, and wetlands all over the farm. I imagine our children carrying on the work here in some way. I want us to enjoy the life we have built after all the struggles. We still have a wall and vegetable garden to finish, and I would love a front porch to put all the boots and coats in. Then this farmhouse is kind of done. It has been a long journey, but it is now time for us to do more than survive, more than just get through the days.

We are kinder to each other now, because we understand ourselves better. We are happiest when together, whether that is tidying up the kitchen or sheep shed, playing with the kids, working outside on the farm or inside on a piece of writing. We know our limits better too. We spent too many years pushing ourselves and each other beyond them. A swim in the lake is as important as making the hay – both have to be given time for a good life.

Floss has been patient while I take in the view, but she wants to be on the move again. We walk up the road and turn right into the old droving lane that runs behind our house. She potters amongst fallen tree branches and starts to stalk along the fence when she sees the sheep in our neighbour's field. The ewes are settled with their lambs and barely notice her. She pauses to sniff where a badger has left its scent and then trots along in front of me. The sun is dipping and will soon be behind the mountain of Blencathra in the distance, but for now it is casting golden rays of light over the whole valley. It is the most magical time of the day. A curlew calls to its mate in the sky above me. They are flying back to their roosting place on the hillside for the night. Their calls, to me, always sounds like they are singing to each other: 'I'm here, it's OK, come and fly alongside me.' It is one of the most beautiful sounds on earth.

My feet walk the ancient path a hundred shepherds have walked, taking their flocks to and from the fell. Twenty minutes ago I was caught up in the slog of the daily chores. It seemed easy then to think I had achieved very little with my day when it consisted of washing, tidying up, school runs, cooking and an hour or two of farm work.

The sun is behind the mountain now. I pull my jumper on and stand for a moment longer before I head back to the house. I feel completely serene. I feel at one with this place I call home. I love the life, with all its ups and downs, that I live here in this family, the family that James and I created.

These days with all six of us here together in this farmhouse are so precious. They are the best days of my life.

The chores always seem mundane, like things we should be escaping from, but what is a life without work? It becomes empty and meaningless. Doing things like cooking, cleaning and errands to help my family is purposeful and important to me. Caring roles in our society are all too often invisible, ignored in the crazy 'look at me' world we live in. Individuals are celebrated instead of families and communities. But no one does anything entirely on their own. I know lots of brilliant people, but there is always a team of hidden, hardworking souls around them making it all possible. I know that the small domestic things matter; they always have – my mum taught me that. Learning that the word 'mundane' has its roots in the Latin word 'mundanus', *of the world*, made me see everything through a different lens. To me, caring for my family is, and always has been, the most important work in the world.

# HELPFUL LISTS

# PANTRY STAPLES

## BASIC LONG-LASTING STAPLES

Flour: self-raising and plain, strong white bread flour and wholemeal flour

Sugar: granulated, caster, soft brown and dark brown sugar

Bread: naan, wraps, rolls or sliced

A few jars of good-quality cook-in sauces: butter chicken curry sauce,
sweet and sour, or tomato and basil

Curry paste: korma, tikka, massaman and green or red Thai paste

Tins of coconut milk

Tins of tomatoes: chopped or whole plum tomatoes

Passata (blended tomato sauce)

Tuna: tinned in spring water

Sweetcorn: tinned

Beans: tins of baked beans in tomato sauce, and cans or jars of kidney
beans, borlotti beans, butter beans and chickpeas

Ready-cooked lentils: red or green in packs, ready to add to meals

Dried lentils/couscous

Dried noodles: egg or rice noodles or packs of 'straight to wok' noodles

Pasta: I have a selection of spaghetti, linguine, tagliatelle, macaroni,
penne, fusilli and rigatoni

Rice: basmati, arborio (risotto rice), pudding rice and wild rice

Rolled oats

Dark chocolate (several blocks hidden in my pantry)

Syrup: golden and maple

Treacle

Tinned fruit: pears, pineapple and peaches

Packs of fruit-flavoured jelly

Spreads: jam, chocolate spread and peanut butter

Condiments: mint sauce, horseradish, cranberry sauce, chutney, relish

Vinegar: malt, red or white wine, and balsamic

Oil: olive and groundnut

Organic stock cubes: chicken, beef and vegetable

## FRIDGE STAPLES

Butter: salted and unsalted

Cream: double cream and full-fat crème fraiche

Cheese: a selection of Parmesan, Cheddar and something like
 Wensleydale or Red Leicester

Yogurt: natural or Greek-style

Meat: cuts that I am defrosting, as well as lardons/diced pancetta, bacon,
 minced lamb or beef, chicken breasts or thighs, and cold cooked
 ham, beef or turkey

Potatoes, sweet potatoes, carrots, onions, garlic

Vegetables in season

Fruits in season

## FREEZER STAPLES

Frozen peas

Frozen green beans

Oven chips (ones made with beef dripping)

Meat: a variety of whole joints of chicken, lamb, pork and beef, plus
 sausages, burgers, steaks, diced casserole and mince

Fish fingers

Prawns

Bread

Pizza/garlic bread

Pastry: packs of shortcrust and puff pastry

Berries: blueberries, raspberries and brambles

Ice cream

# MEALS WE SHOULD NEVER FEEL GUILTY ABOUT

Most of my meals are quick. I rush into the house from jobs outside or picking the kids up and often don't know what we're going to eat. Here are a few ideas for putting food on the table in a really short time. The main thing here is that we eat together and there is very little thinking involved.

## A COLD SUPPER

Cured meats, hummus, bread and cheese, with some raw veg like cherry tomatoes, carrot and cucumber sticks, and a few boiled eggs sliced in half – this is a meal. Don't feel guilty for not cooking. We call it a 'platter' in our house. It's various 'fridge bits' unwrapped and put out so everyone can eat what they like best.

## ANYTHING ON TOAST

For me, it is mostly baked beans and scrambled or fried egg. I also like a few mushrooms or a tomato and some bacon.

## SOUP

Ideally home-made and stored in the fridge or freezer. But you can make a very simple vegetable soup in a pan or microwave in very little time. To turn it into a more filling meal, make it with good stock and add toppings like pumpkin seeds, bacon or crispy fried onions, or small pieces of chicken or salmon. Or have cold meat or cheese and bread on the side.

## FISH FINGER SANDWICHES

These can be as basic as a slice of bread and a couple of fish fingers, or make them fancy by toasting bread rolls and adding mayonnaise or tartar sauce and salad leaves.

## PIZZA

A simple margherita pizza combines all the food groups: starchy carbs, vegetable sauce, dairy and protein toppings. When you can, try to make the pizzas yourself. Bases can be batch-made and stored in the fridge for two days or in the freezer between layers of greaseproof paper – just stretch them out a little again before using. Make the sauce yourself by roasting a tray of roughly chopped tomatoes, a couple of cloves of garlic and a chopped onion, drizzled in olive oil and seasoned well. Blitz to form a sauce (or use a simple jar of tomato passata), and top with good-quality mozzarella or other grated cheese. Add any toppings you like but don't overload the pizza as this will make it soggy. Or just buy the best-quality pizza you can afford. Check the ingredients list on the pack – you're looking for a simple list, as far as possible additive- and preservative-free.

## BACON

Most meals can be improved by adding a little bacon. Choosing the right bacon starts with knowing how the pig was raised and what it was fed on. Pigs raised outdoors, grown slowly and organically, have a richer, meatier, more distinct flavour. Meat like this is nutrient-dense. A little goes a long way.

Pork that has been raised indoors with very little room to move around and fed grain so as to grow quickly is obviously much cheaper on the supermarket shelves than the above, but you get what you pay for. Meat like this is an inferior product, usually watery and bland. And – crucially, under the current free-trade deals our government is pursuing – we must be careful to check where it has been produced, because we could be buying meat that has been farmed in terrible farming systems and shipped across the globe. I always advocate choosing British produce and buying the best you can afford.

Bacon can be used in a number of ways. We like it thickly sliced, fried in a heavy pan and served with egg on toast or in a bread roll. I use pancetta, cubes of fatty cured bacon, regularly too, in pasta sauces, chicken pies, bolognese sauce, beef stews or on top of salads or soups. Pancetta or a little fried bacon makes a bowl of boiled new potatoes or sautéed leeks delicious when scattered on top and drizzled with the fat from the pan. The kids often make pancakes with crispy streaky bacon for an American-style breakfast.

### SMOKED VS UNSMOKED

Most supermarket smoked bacon is soaked in a chemical brine or injected with a smoke-flavoured additive. If I am buying bacon I prefer unsmoked, knowing that my bacon has been cured simply with salt and sugar or spices. (Brown sugar or maple syrup with the curing salt makes a good mix to rub over the pork belly.) It will have been left to rest for a week or more before slicing. Traditional methods of smoking bacon over woodchips are still used, but it just takes a little time and effort to source good-quality produce.

## CREATIVE LUNCHBOX IDEAS

For a change to the regular sandwich, be creative with different bread types: wraps, ciabatta, focaccia, sourdough, bake-at-home baguettes, malted rolls and rye bread are all a change from ultra-processed supermarket loaves.

A FEW IDEAS TO MAKE LUNCHTIME
MORE INTERESTING:

Sourdough toasted, rubbed with raw garlic and topped bruschetta-style with fresh tomatoes, virgin olive oil, salt and pepper

Soft sub rolls filled with egg mayonnaise and chopped salad including cherry tomatoes, iceberg lettuce, grapes and peppers

Farmhouse loaf with ham, mustard, tomatoes and pickle, or roast beef, horseradish and lettuce

Rye bread with scrambled egg or sliced boiled eggs and smoked salmon

Hot baguettes with minute steaks, mushrooms and onions

Ciabatta rolls with pulled pork and barbecue or apple sauce

Wraps filled with crispy chicken, paprika mayonnaise and shredded slaw and salad, or falafels, hummus and raw carrot sticks

Flatbreads with slow-cooked pulled lamb, salad, tzatziki and pomegranate seeds

Toasted sandwiches with bacon, Brie and cranberry sauce

Baguettes or rolls filled with Cumberland sausages with red onion marmalade

Malted bread filled with sliced cold roast chicken with mango chutney, mayonnaise and lettuce

The ultimate Christmas sandwich is leftover roast turkey in a white bread roll, spread with stuffing and cranberry sauce, and a simple coleslaw and salty crisps on the side

# SIMPLE STUDENT MEALS

Student food shouldn't just be pasta or Pot Noodles. There are so many easy things you can make that only take a little planning and shopping. Simple meals don't have to be expensive, but remember that the 'cost' should take into account investing in nutrient-dense foods that satiate and stop you wanting to buy and snack on processed foods afterwards.

## SUPPER IN A ROASTING TRAY

Place chicken thighs and legs, chunks of sweet potato and any other veg you like in a roasting tray – pepper, courgette, squash, onion, tomato. Scatter with herbs, drizzle with olive oil and season well with salt and pepper. Roast at 200°C for 30 minutes until the chicken is cooked through. Any leftovers from the tray of cooked roasted vegetables are great for mixing with pasta and passata or using up in a soup.

## SOUP

Sauté a chopped onion and some vegetables in olive oil to release their flavour before adding stock, and season well. Blend if you like a smooth soup, and add milk or cream or a tin of tomatoes for a more filling meal.

## MACARONI CHEESE WITH PANCETTA AND LEEKS

Make a basic white sauce (see p. 262–3, without adding the chicken stock cube), and add some grated cheese and drained cooked macaroni. In a frying pan sizzle up some pancetta, and meanwhile I like to cook my leeks in the microwave for 5–8 minutes with a knob of butter and a splash of water to keep them bright green. Stir the pasta and sauce together with the pancetta and leeks, top with breadcrumbs or grated cheese, and finish under the grill or in the oven.

## VEGETABLE STIR-FRIED NOODLES

Serve with sliced chicken breast, sliced minute steak or a few prawns, and any seasonings (soy, garlic, lemongrass, ginger).

### SAUSAGE CASSEROLE OR SAUSAGE GNOCCHI

Cook the sausages first, slice up and mix with a tomato and vegetable sauce, adding in a pack of pre-cooked tomatoey lentils if you like, then stir in a pack of gnocchi (cooked to packet instructions) or serve with mash.

### QUICK BOLOGNESE

Brown the mince in onion and garlic. Season well, add in chopped tomatoes or passata and cook for 30–40 minutes. Make it into a chilli by adding some chopped peppers, seasoning with chilli powder and cumin before adding the tomatoes and then adding a tin of red kidney beans 10 minutes before the end.

### JACKET POTATOES WITH DIFFERENT FILLINGS

Cheese and apple, tuna mayonnaise, baked beans, cooked lentils and bacon, leftover bolognese, chicken curry or sausage casserole.

### QUICK CURRY WITH CHICKEN OR VEG

Pan-fry your chicken or veg until cooked through or soft, add 2 tbsp of curry paste and sizzle for 2 minutes to release the flavours, then add a tin of coconut milk, bring to a simmer and add some greens like spinach or shredded kale for 2 minutes at the end.

### SALAD BOWLS

Cheese, chicken, tuna or prawns with shredded lettuce, diced vegetables, boiled eggs and a tasty dressing.

### A SIMPLE STROGANOFF

Pan-fry 1 or 2 pieces of steak or a sliced pork chop with onions, garlic and mushrooms, a little Dijon mustard, white wine or chicken stock and a tub of single cream. Serve with pasta or rice.

### AN EASY RISOTTO

With onion, garlic, stock and risotto rice (following the packet cooking instructions) and any flavours – frozen prawns and peas, mushroom and chicken, leek and pancetta or butternut squash and toasted pine nuts.

Students, don't forget that a piece of fish, lamb or pork with some boiled, mashed or roasted potatoes and a few steamed vegetables is a delicious meal. Food should not be about whether it is Instagrammable or from the latest recipe trend. Food is about nourishment and looking after yourself.

## FAVOURITE COOKBOOKS

I like looking at cookery books for inspiration and presentation ideas, but the bulk of my cooking is done by instinct and either buying fresh seasonal ingredients or using up what I already have in the freezer, basing my meals around good-quality meat. I don't want a recipe that calls for lots of things I don't have so I tend to avoid cooking from a book. I have learned how to cook from trial and error, watching and copying, substituting ingredients according to my family's tastes, and experimenting.

I learned basic techniques like roasting, sautéing, frying, boiling, braising and baking by practising over time. I've honed extra techniques from watching chefs on TV or reading books. I prefer not to have a photograph of what I am making because the real thing never looks the same as the staged food.

*Leith's Cookery Bible* by Prue Leith and Caroline Waldegrave
*Salt, Fat, Acid, Heat* by Samin Nosrat
*How to Eat* by Nigella Lawson
*Midnight Chicken* and *The Year of Miracles* by Ella Risbridger
*At Elizabeth David's Table* compiled by Jill Norman
*Delia's Frugal Food* by Delia Smith

## NOURISHING FOOD FOR SELF-CARE

Thick toast, scrambled eggs, smoked salmon

Baked potato with butter, baked beans and good-quality thick-sliced ham

Mashed potato with butter and cream and sausages with onion gravy and
peas

Butter chicken curry, rice and naan bread

Ham hock broth (see p. 248)

Chicken broth with noodles and vegetables


## MEALS FOR WHEN I'M IN SURVIVAL MODE

Toast with marmalade and a boiled egg, a slice of thick brown bread with peanut butter and an apple, or a bowl of cereal gets me through if I am not able to cook. At busy times I rely on simple meals pieced together from things I always have, like sausages, burgers, bacon, chicken and beef mince in the freezer.

Rolls filled with beef, lamb or pork burgers, sausages or bacon – basically
anything in a bread roll that is quick and effortless

Deli lunch: pies, Scotch eggs, ham and beef and ready-made salads with
bread

Pasta with meatballs and passata

Cheat's bolognese: beef mince and a ready-made tomato sauce with
pasta

Cheat's curry: find a sauce you like – we like butter chicken – and then
pan-fry chicken and add the sauce, served with rice

Tins of soup

Baked beans on toast

## EASY WAYS TO FEED LITTLE ONES

Cheese and crackers

Breadsticks and dips: hummus, cream cheese or a smoked-salmon-pâté-style dip (smoked salmon blended with cream cheese and lemon juice)

Raw veg sticks – carrot, cucumber and celery – and cherry tomatoes

Cooked pasta twists

Toast fingers with cheese

Baguette and butter

Boiled eggs, halved

Any kind of bread and a filling your child likes: ham, beef, chicken, tuna, salmon, cheese or egg mayo

Cold sausages, sliced up

Soup, home-made and kept warm in a flask

Yogurt

Fruits: blueberries, raspberries, strawberries and halved grapes and little oranges or a banana

Home-made cakes like muffins, tea bread, cookies and flapjacks

## A NOTE ON BUYING CAKE

Most bought cakes contain a ton of preservatives and added raising agents. A long shelf life also indicates that the cake is full of unknown substances. Look for a bakery that makes them on site and serves them fresh.

# EXTRA RECIPES

# LEMON DRIZZLE

Prep 15 minutes

Cook 30 minutes

Makes 1 lemon drizzle loaf cake

## Ingredients

100g/4oz unsalted butter, softened to room temperature

100g/4oz caster sugar

2 eggs

225g/8oz self-raising flour

1 lemon, zest and juice (reserve the juice for the topping)

2 tbsp milk

1 tbsp sugar for topping (or 3 or 4 tbsp icing sugar)

## Method

1. Heat the oven to 180°C/fan 160°C/gas 4. Grease and line a 1lb loaf tin.
2. Cream the butter and sugar in a mixer or by hand really well until light and fluffy – it can take a good 5 minutes before the sugar dissolves.
3. Add the eggs and a tablespoon of the flour and beat in well.
4. Add the rest of the flour and fold in gently with a metal spoon to keep the mixture light.
5. Stir in the lemon zest and milk. You will have quite a firm batter. Spoon the batter into the prepared loaf tin, and gently smooth the top.
6. Bake for 30 minutes until golden and a skewer or cocktail stick inserted into the centre of the cake comes out clean. Leave to cool in the tin while you make the topping.
7. In a small pan, heat the lemon juice and sugar until the sugar has dissolved.
8. Prick the warm loaf with a skewer and pour over the hot topping. Alternatively, leave to cool and mix the lemon juice with 3 or 4 tbsp icing sugar and drizzle over the top of the cake for a runny icing.

Note: When baking, it is always worth doubling the recipe up and making two cakes, one to eat and one to give away or freeze for another occasion.

## GINGERBREAD

**Prep 20 minutes**
**Cook 1 hour**

### Ingredients

225g/8oz butter
225g/8oz soft dark brown sugar
100g/4oz treacle
100g/4oz golden syrup
340g/12oz plain flour
2 tsp ground ginger
1 tbsp ground cinnamon
280ml/½ pint whole milk
2 tsp bicarbonate of soda
2 eggs, beaten

### Method

1. Heat the oven to 180°C/fan 160°C/gas 4. Grease and line a deep 20 x 30cm roasting tin.
2. Melt the butter, sugar, treacle and syrup in a pan without boiling. Set to cool. When I measure my syrup and treacle with a tablespoon I grease the spoon first, and it slips off easily straight onto the scales.
3. Sift the flour with the ginger and cinnamon into a large bowl or stand mixer.
4. Warm the milk in another pan, or microwave gently for 30 seconds, and stir in the bicarbonate of soda.
5. Pour the cooled mixture from the first pan into the flour and spices with the mixer (if using) set on low, then add the beaten eggs and the warm milk to form a loose batter. Keep scraping the bottom of the bowl to incorporate any mixture that has got stuck.

6. Bake in the centre of the oven for 1 hour, covering the top with foil or greaseproof paper after about 40 minutes.

7. It is ready when a skewer comes out clean. Leave to cool before turning out and cutting into squares.

Note: Gingerbread forms a sticky top if left in a tin for a few days and improves. It also freezes well so I often make two at a time.

## COFFEE CAKE

**Prep 45 minutes**

**Cook 20–25 minutes**

Serves 8

Ingredients

For the cake:

170g/6oz unsalted butter (room temperature)

170g/6oz soft light brown sugar

3 eggs

170g/6oz self-raising flour

2 tbsp strong coffee (I mix 6 tsp of instant coffee into 5 or 6 tbsp of boiling water for the cake, filling and topping)

For the buttercream filling:

50g/2oz softened unsalted butter

100g/4oz sifted icing sugar

1 tbsp strong coffee

1 tbsp whole milk

For the coffee fudge frosting:

170g/6oz icing sugar

25g/1oz unsalted butter

50g/2oz soft brown sugar

2 tbsp single cream

2 tbsp strong coffee

walnut halves, to decorate (optional)

## Method

1. Heat the oven to 180°C/fan 160°C/gas 4. Grease and line 2 x 20cm sandwich tins. For the cake: in a large mixing bowl, or stand mixer, whisk the butter and sugar to a creamy, smooth consistency, scraping down the bowl to make sure it is all combined.

2. Add in one egg at a time with a little of the flour to stop the mix curdling, then fold in the rest of the flour.

3. Beat in the strong coffee until well combined. Scrape the cake batter out of the bowl with a spatula into the two lined cake tins, denting the centre of the mixture to help them rise evenly in the oven.

4. Bake in the middle of the oven for 25–30 minutes but check them after 20 minutes. The cakes are ready when you can insert a skewer into the centre of the cake (I use a knitting needle) and it comes out clean, with no raw cake mix on it. They will also be golden and springy to the touch. Cool for 10 minutes in their tins then carefully tip them onto a rack to cool.

5. For the filling: whisk the soft butter first before adding the icing sugar or you will end up with clouds of icing sugar as you try to break down the butter. I put a tea towel over my bowl and use a simple electric hand whisk.

6. Add the coffee and milk and whisk thoroughly until smooth. Cover, and leave to set in the fridge.

7. When the cakes are completely cool, turn one out onto a plate, gently spread with the cooled buttercream and then place the second cake on top.

8. To make the topping, sift the icing sugar into a heatproof bowl.

9. Put all the other ingredients into a pan (butter, sugar, cream and coffee) and heat, keeping a close eye so you don't burn it, until it begins to boil and starts to rise up in the pan. Remove the pan from the heat and pour the hot mixture steadily into the icing sugar. Using a balloon whisk, beat the mixture well until smooth, spreadable and fudge-like. This topping doubles easily and keeps well, covered and stored in the fridge or freezer, ready for icing another cake.

10. Top the cake with the fudge frosting while it's still warm, spreading it to the edges of the cake with a metal palette knife – if it has gone too hard to spread, just gently warm it in a microwave or a pan until it is spreadable again. Decorate the top with a few walnut halves if you like.

## DROP SCONES

**Prep 15 minutes**

**Cook 15 minutes** (if cooking in batches)

Makes 6 large or 10 small scones

### Ingredients

110g/4oz self-raising flour

a pinch of salt

15g/½oz unsalted butter, plus extra for frying

50g/2oz caster sugar

1 large egg

4 tbsp milk

butter and jam to serve

### Method

1. Sift the flour into a bowl and add the salt.
2. Rub the butter into the flour with your fingertips. (If doing a double batch, I would use a food processor to rub in the flour.)
3. Add the sugar to the bowl and stir.
4. Whisk the egg and milk together in a bowl or small jug, then pour into the flour, mixing well to form a smooth batter. The mixture will be quite thick.
5. Heat a small knob of butter in a frying pan/flat griddle pan, swirling it so it covers the base. Carefully drop a tablespoon of the scone batter onto the hot pan, scraping the mix off the spoon with your finger or another spoon, with plenty of space between scones, as they spread whilst cooking.

6. Wait for little bubbles to appear in the scones – after roughly 3 minutes – before turning them, managing the heat of the pan carefully so you don't burn them.
7. Remove to a plate to serve. Best eaten warm, spread with butter and jam.

## SCONES

Prep 30 minutes

Cook 10 minutes

Makes 6 large scones

Ingredients

340g/12oz self-raising flour

¼ tsp fine sea salt

75g/3oz unsalted butter

3 tbsp caster sugar

100g/4oz sultanas

175ml/6fl oz milk

¼ of a lemon

½ tsp vanilla extract

1 egg, beaten, to glaze (or extra milk)

Method

1. Heat the oven to 190°C/fan 170°C/gas 5. Put a baking sheet in the oven to heat up.
2. In a large bowl, rub the flour, salt and butter together to a breadcrumb consistency, then stir in the sugar with the sultanas.
3. Warm the milk in a pan, or in the microwave for 30 seconds, then stir in the juice from the lemon, and the vanilla.
4. Add the wet ingredients to the dry – with a fork or a knife initially, to stop your fingers from becoming too sticky – then bring it together to form a dough. Use your hands when it is in a rough ball. Don't overwork it.
5. Lightly press out or roll the dough on a floured worktop to roughly 4cm deep. Use a cutter (mine is 7cm) to cut out 6 scones.

6. Bring together the scraps of dough after the first scones are cut out and pat with your hand and cut out more until you have no dough left.
7. Brush the tops with beaten egg (or milk) and lift onto the preheated baking sheet.
8. Bake for 8–10 minutes until risen and golden brown. Allow to cool.

## FANCY NIBBLES

### SMOKED SALMON TRIANGLES

Spread good-quality brown bread with softened butter or cream cheese and top with thin slices of smoked salmon. Cut the crusts off and cut into triangles, grind over some black pepper and serve with wedges of lemon.

### HUMMUS WITH PARMA HAM

Cut a baguette into thin slices and toast, or use crostini, and spread with hummus and top with curled pieces of Parma ham.

For a quick home-made hummus (makes 340g/12oz): Blend 1 x 400g can of chickpeas (drained and washed) with ½ garlic clove, 2 tbsp olive oil, juice of ½ large lemon, 2 tbsp tahini paste and 1 tbsp water. Season generously with salt and pepper and loosen with more water as required. Season with extra lemon juice, salt or tahini.

### PARMA HAM WITH ASPARAGUS OR MELON

Steam or parboil spears of asparagus, leave to cool then wrap with Parma ham. Alternatively, wrap chunks of melon in the thin ham.

### CROSTINI

On a baking tray, arrange thin slices of baguette or ciabatta loaf – 3–4 per person – drizzle well with olive oil and toast in the oven at 190°C/fan 170°C/gas 5 for 5–10 minutes, turning once. Leave to cool and serve with the hummus or either of the following pâtés.

## CHICKEN LIVER PÂTÉ

Makes 700g/1.5lb: 1 small loaf, or 6 portions in ramekins

### Ingredients

450g/1lb chicken livers, membranes removed (you can pre-soak the
livers in milk to help remove the membrane)

200g/7oz melted butter

50ml/2fl oz double cream

1 tbsp brandy

### Method

1. Sauté the chicken livers in a little of the melted butter for 2–3 minutes and leave to rest.
2. Transfer to a blender and whizz until smooth.
3. Add the rest of the melted butter and the cream, and season well with salt and pepper. Blitz again until completely smooth.
4. Add the brandy and give a quick blend before scraping out into either several small ramekins or a 700g/1.5lb loaf tin lined with cling film. Tap the tin or ramekins on the kitchen surface to remove any air bubbles in the pâté.
5. Chill in the fridge, ideally overnight.
6. Serve this pâté spread on toast with some sliced cherry tomatoes, gherkins or sliced red onions marinated in red wine vinegar.

## SMOKED SALMON PÂTÉ

Makes 200g/7oz

### Ingredients

150g/5oz pack of smoked salmon

2 tbsp full-fat crème fraiche

2 tbsp cream cheese

juice of ½ lemon

1 tsp fresh grated horseradish or 1 tbsp horseradish cream (optional)

capers, parsley or dill to serve

## Method

1. Place all ingredients, apart from any toppings you've chosen, in a food processor and blitz to make a slightly coarse pâté – not a puree.
2. Scrape out into a bowl to check the seasoning, adding a grind or two of black pepper. Taste before adding any salt as smoked salmon is already salty.
3. Serve in a bowl, topped with a few capers, finely chopped fresh parsley or dill.

## OSMAN'S SPICY EGGS

**Prep 15 minutes**

**Cook 20 minutes**

Serves 1–2

### Ingredients

1 medium onion, finely sliced

1 tbsp oil

1–2 green chillies, finely chopped, or to your taste

¼ tsp chilli flakes

¼ tsp chilli powder

¼ tsp salt

1 small tomato, chopped

3 eggs, beaten

30g/1oz coriander, chopped

naan, chapati, flatbreads to serve

yogurt to serve

### Method

1. Sauté the onion in the oil for 10 minutes until soft.
2. Add the chopped chilli, spices, salt and tomato, and cook for 2–3 minutes.
3. Add the eggs and let them sit for a minute before stirring gently with a spatula until cooked to your liking.

4. Stir through the coriander and warm through.
5. Serve with warm naan bread, chapati or flatbreads, and yogurt.

## ŁUKASZ' SHAKSHUKA

**Prep 20 minutes**

**Cook 40 minutes**

Serves 2

### Ingredients

1 onion, finely sliced

2 cloves of garlic, minced

3 bell peppers (red, green or yellow), halved, deseeded and topped then finely sliced

½ courgette, finely diced

6 fresh tomatoes, chopped, or 400g can chopped tomatoes or passata

½ tsp sugar

1 tsp hot smoked paprika

4 eggs

coriander or parsley to serve

### Method

1. Heat the oven to 200°C/fan 180°C/gas 6. Heat a little olive oil in a wide, heavy-bottomed, ovenproof frying pan.
2. Sauté the onion and garlic for 2–3 minutes and add the peppers and courgette. Cook for another 15 minutes, stirring gently to soften the vegetables.
3. Add the tomatoes, some salt and pepper, sugar and paprika.
4. Cook for a further 10 minutes on a low heat with a lid on.
5. Make a small well at the edge of the pan and crack an egg in.
6. Repeat for all 4 eggs and then put the pan, uncovered, in the oven for 7–9 minutes until the eggs are just cooked. This is best if the eggs are still runny.
7. Scatter with fresh herbs and serve.

# ACKNOWLEDGEMENTS

In 2020 life as we knew it paused because of Covid. For me, it was the first time I could really take stock of the world around me, to stop chasing around 'doing' and to look closely at the life we had created on the farm. I had time to think about where I had come from, and why I did the things I did. It was a chance to focus on myself for the first time in many years and I used writing as a way to express my thoughts and ideas.

It has been an intense experience to write such a personal memoir. It took me on a journey I never imagined I would take. I started by writing scenes – flashbacks of growing up watching Mum and Grandma making marmalade, or learning to cook meals from things already in our pantry. I shared the pages with James, and he loved my writing. He encouraged me to keep going and has been my cheerleader, my sounding board, my first reader. I could not have done this without his unshakeable belief in me. To pour myself onto the page has been messy and raw, revealing, joyful and hard at times, but ultimately we are stronger because of it. We have gone through a shift in consciousness, not only about ourselves and our own roles but about each other's too.

Thank you, James. I love you and I love our life together.

Farms, families and books all need a village of people to help and support them. I am lucky to have such a brilliant village.

Jim Gill, my agent, thank you for all those early conversations we had about sausage rolls. Thank you for your care and attention to detail and for getting my proposal directly under the eyes of the right editors. And thanks to the whole team at United Agents who do the unseen work to make things happen.

Thank you to the editors that showed interest in my writing at the very start: Chloe Currens, Helen Conford, Ione Walder, Lindsey Evans, Liz Gough, Lisa Highton, Amandeep Singh, Adam Humphrey and Vicky Eribo. Your generous words have stayed with me and carried me through the two years of writing this book.

I'm so grateful to Louisa Joyner, my editor, for commissioning my book. Thank you for really seeing me in my early words and giving me the freedom to write the book I had inside me. Thank you for helping me shape it into something I am hugely proud of. Thank you to Eleanor Crow for your attention to detail in the exquisite illustrations that make my words come to life. Your illustration of Bramble still makes me cry. Thank you to Kate Ward for making it look beautiful on the page. I loved showing you around our farm and standing in the meadows and witnessing your reaction. Thank you to Silvia Crompton, for your excellent copy-editing skills and kind words about my writing. Rosie Ramsden, thank you for weighing, measuring and cooking all the food in this book. I don't feel so daunted about sharing my recipes now! And a huge thank you to Mary Cannam and the team at Faber for making me feel so welcome and for working with so much care on the book. It has been a joy to work with you all.

Since I started sharing a window into life on our farm on Instagram, I have had heaps of love, support and encouragement. I have had conversations with strangers who feel like friends – you know who you are. Thank you, I am really grateful.

Thank you to Claire Jeannerat for being my friend through some pretty long days and for encouraging me to start a project of my own. My trip to Switzerland to visit you in 2019 and all our conversations were instrumental in me writing this book. Thank you to Maggie Learmonth for being my creative wise owl, always ready to listen, and for giving me a quiet place to write at very short notice when the wheels were falling off! Nick Offerman, Maria Benjamin,

Łukasz and Joanna Długowska, Sarah Langford, Sophie Gregory and Lindi Phillips – thank you for your friendship and unwavering support and encouragement. Thank you to Elizabeth Wainwright for being genuinely lovely and for reading a first proof so carefully and thoughtfully. Thank you to Susan Ursell, a farmer's wife, who I shared this project with in its early stages around her kitchen table. I always remember your tears, because it made me realise how unseen many of us feel. Thank you to Nicola Estill, Julie Bowman, Helen Gardiner and Angela Barton for being my best and oldest friends. I love that we instantly pick up where we left off and I know you are always there for me.

Thank you to Heather Glennon, Libby Mason and Michelle Brunskill: at different times you have been the heroes of my life, helping look after the kids, cleaning the house and doing the paperwork. I couldn't have done it without you. Jill Rebanks, for being 'the farmer's wife' before me – thank you for always lending a hand, being calm and understanding when life gets hectic.

Thank you to Mum and Dad, Stuart and Alison. I know this is all so personal, and we don't tend to share so much, but sometimes it's important to. Thank you for everything.

My original idea for this book was to tell my story and gather up some recipes as a kind of legacy for my children. Molly, Bea, Isaac and Tom, thank you for bearing with me over many months of writing this book. You have all stepped up in helping out when needed. And suffered slightly too many meals of butter chicken.

Being your mum is my proudest achievement. I love you.

Lastly, to my mum, thank you for being you. You showed me the way, you've always given your family everything you've got, and still held on to who you are. I love you so much for that.